CHEMISTRY RESEARCH AND APPLICATIONS

CRYSTAL VIOLET

PRODUCTION, APPLICATIONS AND PRECAUTIONS

CHEMISTRY RESEARCH AND APPLICATIONS

Additional books and e-books in this series can be found on Nova's website under the Series tab.

CHEMISTRY RESEARCH AND APPLICATIONS

CRYSTAL VIOLET

PRODUCTION, APPLICATIONS AND PRECAUTIONS

VICTOR DUFFET
EDITOR

Copyright © 2019 by Nova Science Publishers, Inc.

All rights reserved. No part of this book may be reproduced, stored in a retrieval system or transmitted in any form or by any means: electronic, electrostatic, magnetic, tape, mechanical photocopying, recording or otherwise without the written permission of the Publisher.

We have partnered with Copyright Clearance Center to make it easy for you to obtain permissions to reuse content from this publication. Simply navigate to this publication's page on Nova's website and locate the "Get Permission" button below the title description. This button is linked directly to the title's permission page on copyright.com. Alternatively, you can visit copyright.com and search by title, ISBN, or ISSN.

For further questions about using the service on copyright.com, please contact:
Copyright Clearance Center
Phone: +1-(978) 750-8400   Fax: +1-(978) 750-4470   E-mail: info@copyright.com.

NOTICE TO THE READER

The Publisher has taken reasonable care in the preparation of this book, but makes no expressed or implied warranty of any kind and assumes no responsibility for any errors or omissions. No liability is assumed for incidental or consequential damages in connection with or arising out of information contained in this book. The Publisher shall not be liable for any special, consequential, or exemplary damages resulting, in whole or in part, from the readers' use of, or reliance upon, this material. Any parts of this book based on government reports are so indicated and copyright is claimed for those parts to the extent applicable to compilations of such works.

Independent verification should be sought for any data, advice or recommendations contained in this book. In addition, no responsibility is assumed by the Publisher for any injury and/or damage to persons or property arising from any methods, products, instructions, ideas or otherwise contained in this publication.

This publication is designed to provide accurate and authoritative information with regard to the subject matter covered herein. It is sold with the clear understanding that the Publisher is not engaged in rendering legal or any other professional services. If legal or any other expert assistance is required, the services of a competent person should be sought. FROM A DECLARATION OF PARTICIPANTS JOINTLY ADOPTED BY A COMMITTEE OF THE AMERICAN BAR ASSOCIATION AND A COMMITTEE OF PUBLISHERS.

Additional color graphics may be available in the e-book version of this book.

Library of Congress Cataloging-in-Publication Data

ISBN: 978-1-53615-806-9

Published by Nova Science Publishers, Inc. † New York

CONTENTS

Preface		vii
Chapter 1	Aspects Concerning the Hypsochrome Effect and Spectral Remission of Crystal Violet *Elena Bercu, Viorica Vasilache and Vasilica Toma*	1
Chapter 2	Environmentally Safe Biosorbents for Crystal Violet Removal from Wastewater *Preeti Pal, Asmaa Benettayeb and Anjali Pal*	31
Chapter 3	Removal of Crystal Violet Dye from Aqueous Solution Using Ash-Based Adsorbent Materials *Denise A. Fungaro, Suzimara Rovani, Tharcila C. R. Bertolini and Flamarion F. Filho*	75
Index		115
Related Nova Publications		121

PREFACE

Crystal Violet: Production, Applications and Precautions opens by presenting the main factors influencing the metachromatic phenomenon. The hypsochromic effect is due to the symmetry of the molecule providing the common electronic signal around the central carbon atom.

Next, the authors discuss further modification of the adsorbents in an effort to improve their adsorption capacity and make them feasible for use in field applications.

The closing chapter reports on the removal of crystal violet dye from water using surfactant-modified zeolite from coal fly ash, surfactant-modified zeolite from coal bottom ash and nanosilica from sugarcane waste ash.

Chapter 1 - The paper presents the main factors influencing the metachromatic phenomenon. The hypsochromic effect, unexpected in the case of auxochrome group buildup registered at the absorption of the light of the Crystal Violet (CV) is due to the symmetry of the molecule providing the common electronic signal around the central carbon atom. Taking into account that triaminotriphenymethanic colorants present very pronounced phenomena of remission, the effect of two anionic maleic polyelectrolytes (NaM-S, NaM-AV) on Melana fibers dyed with CV was analyzed. It was found that the NaM-S and NaM-AV polyelectrolytes manifest a retarding effect irrespective of the concentration at which they are found in processes of dyeing Melana with CV.

Chapter 2 - Crystal violet (CV) is a triarylmethane dye; widely applied in coloring paper, as a temporary hair colorant, for dyeing cotton, silk, and wool. It is a cationic dye which is also known as gentian violet. CV is used for the manufacture of paints and printing inks; it is also used as a biological staining agent. In animal and veterinary medicine, it is employed as an antibacterial, antifungal, and antiseptic dye. As a result, CV is the waste product of many of the processes which contaminate the water and also traverse through the food chain, leading to biomagnification. CV is carcinogenic and has been classified as a recalcitrant molecule since it is non-biodegradable, and can persist in a variety of environments. CV can easily interact with negatively charged cell membrane surfaces and can enter into cells and concentrate in the cytoplasm. CV can cause moderate eye irritation, painful sensitisation to light, heartbeat increase, vomiting, shock, cyanosis, jaundice, quadriplegia, and tissue necrosis. Therefore, it is essential to remove CV from industrial effluents before it is discarded into water bodies. CV has complex aromatic structure and is resistant toward chemicals, heat and light, and its biodegradation is a slow process. This review article provides the compilation of the information about CV, its classification and toxicity, various treatment methods, and its removal through different environmentally safe adsorbents. There are ample of biosorbents available for CV removal. One of the objectives of this review article is to organize the scattered available information on CV removal through adsorption using biosorbents including chitosan, surfactant-modified chitosan beads, NaOH-modified rice husk, grapefruit peel, coniferous pinus bark powder (CPBP), de-oiled soya, water hyacinth, peanut hull waste biomass, carboxylate-functionalized sugarcane bagasse, etc. The effectiveness of various adsorbents under different process parameters (such as solution pH, initial dye concentration, adsorbent dosage, and temperature) and their comparative adsorption capacity towards CV adsorption has also been presented. This review paper provides the understanding for further modification of the present adsorbents to improve their adsorption capacity and make them feasible for the field applications.

Chapter 3 - Crystal Violet (CV) is widely used for various purposes and enters into the aquatic systems from the effluents of textile, paint, medical and biotechnological industries. A considerable amount of this dye is lost during manufacturing and processing operations. Contaminated wastewater containing CV must be treated before releasing in the environment because it is highly cytotoxic and carcinogenic to mammalian cells, present mitotic poisoning nature and is non-biodegradable being classified as a recalcitrant molecule. This chapter reports the removal of CV dye from water using surfactant-modified zeolite from coal fly ash (MZSF), surfactant-modified zeolite from coal bottom ash (MZSB) and nanosilica from sugarcane waste ash (SiO_2NP). The adsorbent materials were characterized to obtain chemical and mineralogical composition and others physicochemical properties. The adsorption kinetic of CV onto adsorbents was discussed using the pseudo-first order, pseudo-second order, and Elovich models. The Langmuir and Freundlich isotherm models were used to describe the equilibrium adsorption data. The maximum adsorption capacities were 36.7 mg g^{-1} and 21.1 mg g^{-1} for CV/MZSF and CV/MZSB, respectively. The adsorption process of CV/SiO_2NP achieves equilibrium in 60 min of contact time, and the maximum adsorption capacity was 117.98 mg g^{-1}. Application of the adsorbent materials synthesized from agricultural waste and coal combustion products can ensure the sustainability and cost-effectiveness of treating effluent containing CV dye, especially effluent from the textile industries generated in large quantity.

In: Crystal Violet
Editor: Victor Duffet

ISBN: 978-1-53615-806-9
© 2019 Nova Science Publishers, Inc.

Chapter 1

ASPECTS CONCERNING THE HYPSOCHROME EFFECT AND SPECTRAL REMISSION OF CRYSTAL VIOLET

Elena Bercu[1], Viorica Vasilache[2,*] and Vasilica Toma[3]

[1]Metropolitan Church of Moldavia and Bucovina,
Department for Restoration and Preservation of the
Christian art "RESURRECTIO",
Iasi, Romania

[2]"Alexandru Ioan Cuza" University of Iasi, ARHEOINVEST Center,
Institute of Interdisciplinary Research, Romania

[3]"Gr.T. Popa" University of Medicine and Pharmacy,
Faculty of Dental Medicine, Iasi, Romania

ABSTRACT

The paper presents the main factors influencing the metachromatic phenomenon. The hypsochromic effect, unexpected in the case of auxochrome group buildup registered at the absorption of the light of the Crystal Violet (CV) is due to the symmetry of the molecule providing the common electronic signal around the central carbon atom.

---

[*] Corresponding Author's E-mail: viorica_18v@yahoo.com.

Taking into account that triaminotriphenymethanic colorants present very pronounced phenomena of remission, the effect of two anionic maleic polyelectrolytes (NaM-S, NaM-AV) on Melana fibers dyed with CV was analyzed. It was found that the NaM-S and NaM-AV polyelectrolytes manifest a retarding effect irrespective of the concentration at which they are found in processes of dyeing Melana with CV.

Keywords: crystal violet, remission, hypsochromic effect, hydrophilic, hydrophobic

1. INTRODUCTION

Ionic colorants tend to associate in diluted solutions, leading to the formation of dimers and sometimes aggregates and, as the concentration increases, into large aggregates [1-3].

UV-Vis absorption spectroscopy is the most used method for the study of colorant aggregates according to concentration, in the domain (10^{-3} - 10^{-6}M), in which the monomer and dimer are found in equilibrium. In the UV-Vis absorption spectrum this phenomenon occurs as follows: at low concentrations of the colorant in aqueous solutions the α band corresponding to the monomer appears. As the concentration increases, the α band diminishes and a new band appears, designated the β band, corresponding to the dimer. Further on, as the concentration of the colorant in the solution increases, the β and α bands decrease in intensity and a new band γ (or μ) appears, attributed to the colorant aggregates [4-7].

The association of colorants does not constitute simple dimerizations or simple polymer formations: the ions of the colorants in diluted solutions behave as extended hydrophobic groups, since the charge is delocalized across the entire molecule. Thus, the main cause of aggregation is the π interactions between electrons [8-9].

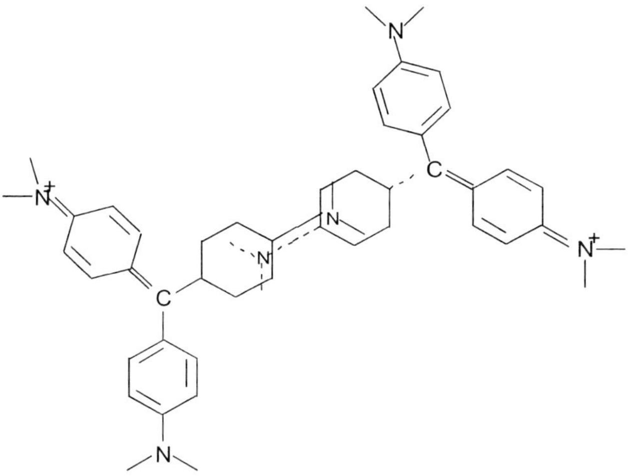

Figure 1. Dimer structure proposed for the ions of Crystal violet in aqueous solutions.

The dimerization of the CV molecules is highly likely due to its flexibility, symmetry and planarity. A dimer structure proposed in the case of Crystal Violet ions in aqueous solution is presented in Figure 1 [10-13].

The London dispersion forces can induce the storing of colorant molecules so that the monomeric units are placed in a sandwich-like fashion at the main macromolecular parallel axes. The increase of the dielectric constant can induce the association of colorant ions, but it can cause a reduction of the coulombian repulsion forces in the area of the colorant ion charge.

The aggregation of the colorants in the solution is highly interesting on account of having applications in multiple domains [1-3, 14-15].

2. AGGREGATION OF COLORANTS IN THE PRESENCE OF POLYELECTROLYTES

By adding anionic polyelectrolytes to the diluted solutions of the cationic colorants, the colorant molecules converge, so that near the charge groups from the surface of the polyelectrolyte occurs the interaction for forming the aggregates similar to those from simple colorant solutions. The theory of storing highlights the fact that in the presence of an excess of polymers, the absorption spectrum of the aggregate colorant returns to the monomeric form of the colorant on account of the redistribution of the colorant's aggregates over the free bonding positions of the polyelectrolyte [16].

M. K. Pal and collabs. have shown that the appearance of the metachromatic phenomenon in the case of the cationic colorants bonded to the polyanions depends on three types of interactions [1-3, 17-19]:

1. electrostatic interaction between the colorant and the polyanions;
2. hydrophobic interactions between the bonded colorants, similar to the formation of the micelles in the aqueous solution of a ionic detergent;
3. interaction between the π electrons of the molecules of adjacent bonded colorants.

Some authors have stated that the main factor responsible for the association of colorants is the energy emitted by overlapping the π electronic clouds that demand the existence of strong hydrophobic interactions [20].

The metachromatic compounds present absorption bands (μ band) at wavelengths of $50 \div 100$ mμ, or smaller than the wavelengths of the α band corresponding to a monomeric colorant dissolved in solution [21].

The spectroscopic studies on the metachromatic behavior of the organic colorants in the presence of polymers have shown that the μ band is not due only to the formation of bonded colorant species, but also to the aggregation of molecules of colorant on the chain of the polyanion. The ordered colorant aggregates from the solutions with high concentrations are reflected in the absorption spectrum in the form of a sharper band. The width and number of metachromatic bands induced by some polyanions correspond to a chaotic aggregation of the colorant cations bonded to the polyanions [1, 17, 22].

It was initially found that metachromasia can be induced in solutions in the absence of chromotrope substances in three different manners: by increasing the concentration of colorants, by adding salt, and by decreasing the values of the dielectric constant of the solvent from the medium. The association of colorants in aqueous solutions and the metachromatic phenomenon are generally found together in the case of colorants that present aromatic systems with partially delocalized charges. The colorants that have charges on aliphatic groups present association constants greater than the colorants with free anionic groups. In such systems, the problems occurring at the dimerization of the colorants and of the nonstoichiometric interactions with opposing charges from the polyelectrolyte are foremost due to the hydrophobic interactions [2, 23].

The presence of an isosbestic point in an absorption spectrum helps describe the reaction between partners as being a simple acido-basic one, excluding the idea of a complicated reaction with multiple equilibria and phenomena. The absence of isosbestic points from the absorption spectrum suggests the existence of a complex equilibrium between monomers, dimers, trimers, etc. [24-26].

Figure 2 presents the schematic model accepted for the metachromatic phenomenon. The colorant ions are attached to the anionic positions from the chromotrope, by electrostatic bonds; being extremely close to each other, so that they undergo effective polymerization [27].

The polymerization of the colorants leads to the splitting of the energy levels in which the smaller energetic transition is prohibited, experimentally observable in the band of short wavelengths. This polymerization of the colorant molecules can involve hydrophobic bonds or colorant–colorant bonds that influence the spectrum of the metachromatic colorants [28].

The interactions between the colorant and the polyelectrolyte in the solution can be characterized by two types of equilibria [2]:

- adsorption of the colorant on the polyelectrolyte;
- association of storing of the colorant bonded to the polyelectrolyte.

The effect of associating the colorant molecules is apparently promoted by the process of bonding due to the colorant molecules *constrained* inside the polyelectrolyte's "phase", and its local concentration is much larger than the stoichiometric one.

The equilibrium constant for the colorant bonded to the polyelectrolyte in the domain of the P/C < 1 values is expressed as follows [29]:

$$K = [\text{complex}]/[\text{free colorant}] \times [\text{free bonding position}]^\alpha,$$

where: α is an empiric parameter.

The equation above expresses the distribution of the free colorant and of the bonded colorant in the solution.

The wavelength of the metachromatic band reflects the storing power of the bonded colorants. The smaller wavelength of the metachromatic band reflects a greater storing power of the colorants [2, 30].

The strength of the interaction between the bonded colorants can be assessed through the storage coefficient [28]:

$$K = e^{-\Delta F/RT},$$

where: ΔF represents the free interaction energy of the pairs of neighboring bonded colorants; the K value was determined according to the Bradley's formula using the following equation:

$$P/C = (1 - F^{1/2})^{-1} + (K - 1)F^{1/2}(1 + F - F^{1/2})(1 - F^{1/2})^{-1},$$

where: F represents the fraction of unstored bonded colorant molecules.

F's values obtained experimentally at different values of P/C can be calculated using the relation:

$$F = (\varepsilon - \varepsilon_1)/(\varepsilon_2 - \varepsilon_1),$$

where: ε_1 represents the molar coefficient of extinction of the α band at values of the P/C ratio at which the storage of the bonded colorant is efficient, ε_2 corresponds to a greater excess of polyanions, and ε to an arbitrary value of the P/C ratio.

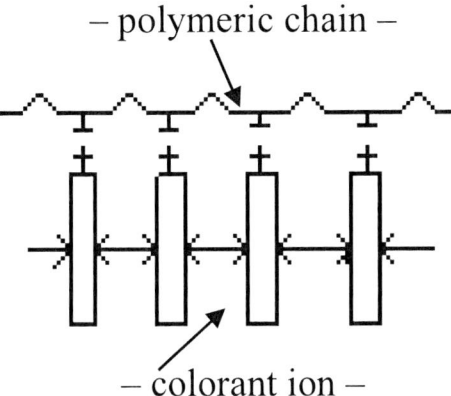

Figure 2. The schematic model for the metachromatic compound that involves colorant–polyanion electrostatic bonds and colorant–colorant hydrophobic bonds.

Since the polyelectrolytes under scrutiny are anionic, and the colorants are cationic, the association of colorant molecules occurs at the centers of charge from the polymer chain; accordingly, the intermolecular repulsion will be diminished, so that the polymer molecule will *coil* on account of the neutralization of the charge density on the chain. This *coiling* is accompanied by a drop in the relative viscosity of the solution.

The polymerization of the colorant induced by the chromotrope polymer (Figure 3) can be explained thus: in a diluted solution the molecule of polyelectrolyte is quite extended, like in diagram A. When the cation of the colorant comes in contact with the anionic group from the chain, the effective charge is partially reduced. When the colorant bonds, the charge is more or less neutralized. The polymer chain coils so that the repulsion force between the negatively-charged groups disappears. This brings the colorant molecules closer (like in case C) and this is more or less similar to the polymerization of the colorant.

The degree of closeness depends on the size of the colorant molecule. If the polymer molecule coils like in diagram D, it is obvious that the colorant molecules 1, 2, 3 cannot occupy the positions indicated because of the stearic hinders. The bonding of the colorant in this case will not be associated to the change in color.

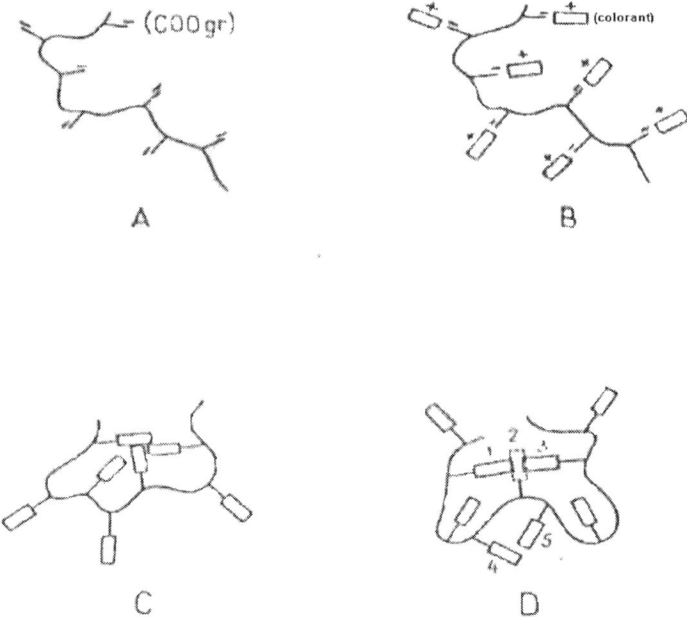

Figure 3. Polymerization of the colorant induced by the chromotrope polymer.

The problems specific to the dimerization of colorants with opposing charges of the polyelectrolytes, to the nonstoichiometry of the interactions in such systems, are mainly due to hydrophobic interactions [1, 2, 31].

2.1. The Direction of Metachromatic Shift

The metachromatic shifts in the absorption spectrum are generally hypsochrome (towards smaller wavelengths), and in some cases we encounter bathochrome metachromatic shifts (towards wider wavelengths) [2, 32].

Studies on metachromatic shifts from the absorption spectrum have shown that the aggregation of the colorant can lead to a hypsochrome or bathochrome shift, and that the direction of shifting is controlled by the arrangement of colorant molecules in the aggregate. If the colorant molecules are arranged in parallel with an angle of inclination $\alpha = 90°$

(stacking), the metachromatic shift is hypsochrome; when the angle of inclination is $\alpha < 90°$ (offsetting), the arrangement leads to a bathochrome shifting (Figure 4) [2, 32, 33].

The bathochromic shifts are less probable than the hypsochromic ones, and the direction of movement depends on the molecular structure of the colorant involved and on the arrangement of ions of colorant in the aggregate.

In accordance with the theory of electron-free position combination applied to the metachromatic phenomenon, the relationship between the position of the metachromatic bands and the orientation of the transition moments of the colorants in the form of oligomer are described thus [2]:

(i) "place near place" parallel orientation of the particular moments of transition that occur in the case of bands with small λ, characterizing monomers;
(ii) "obliquely asymmetric" orientation, occurring in the case of two bands, both at small λ;
(iii) head-tail orientation, caused by a band with large λ.

2.2. Stability of Metachromatic Compounds

The stability of the metachromatic compounds can depend on the simultaneous action of three types of bonds. Each of these three types of bonds can be destroyed by varying a different set of conditions [21].

The electrostatic bond poses particular interest in the case of polyelectrolytes, since they behave as weak electrolytes. This bond has a weak effect on the color of the metachromatic compounds. Moreover, the nature of the polyanion has a weak effect on the position of the μ band.

The second type of bond leads to the formation of micelles, such as those in aqueous solutions of ionic detergents. Because of their planar structure, the cationic colorants can polymerize in parallel planes and probably this is why the bond is responsible for destroying the metachromatic compounds by alcohol or urea [2]. The polymerization caused by this type of bond does not necessarily induce spectral changes.

The third type of bond involved the π electrons of adjacent cationic colorants, and forms with them bonds responsible for the appearance of the μ band of the simple cationic colorant. This bond is probably the most unstable of the three, destroyable by heating the metachromatic compounds, which results in the disappearance of the μ band and the appearance of the α band.

The metachromatic complexes can be destroyed by adding salts, proteins, excessive polyanions, or sufficient quantities of urea of alcohol.

The *stability of the metachromatic compounds* can be *measured* through a number called *value of the middle level*, which depends on the particularities of the colorant and of the polyanion involved [2].

2.3. Factors Influencing the Metachromatic Phenomenon

The metachromatic phenomenon is complex, specific to each pair of partners (colorant–polyelectrolyte), but studies on the metachromatic phenomenon have so far concluded that the main factors influencing this phenomenon are [1-3, 34, 35]:

- the structural parameters of the polyelectrolyte;
- the structural parameters of the colorant;
- the solvent (the pH, ionic strength, charge density, addition of salt);
- the presence or absence of water.

2.3.1. The Structural Parameters of the Polyelectrolyte

Studies on metachromatic reactions have shown that changes in the structural parameters of the polyelectrolytes (length of the chain, charge density, conformation, hydrophobicity, distance between the charge groups along the polyelectrolyte's chain, tacticity, flexibility of the polymer chain, etc.) influence the metachromatic phenomenon [1, 3].

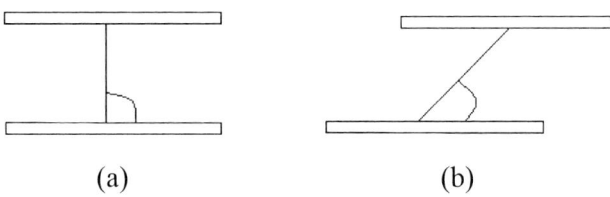

(a)   (b)

Figure 4. The arrangement of colorants as dimers: (a) storing (stacking) with the angle of inclination $\alpha = 90°$; (b) parallel offsetting with an angle of inclination $\alpha < 90°$.

The conformation of a polyelectrolyte molecule, in an aqueous solution, represents the result of a delicate balance between the hydrophilic and hydrophobic interactions. The charged groups from the polyelectrolyte can interact with water, with the ions from the solution, or with each other. On the other hand, the organic remnant of the polyelectrolyte interacts very weakly with water or other elements. This is why corresponding to a strong, fully ionized polyelectrolyte, in solution, is a totally extended conformation, and the distances between the charged groups are maximal. The position and intensity of the absorption band of the bonded colorant species can be influenced by the characteristics of the polymers [29].

The studies concerning the metachromatic interaction of Crystal violet with sodium polyphosphates have shown that by varying the length of the polymeric chain, the intensity of the metachromatic band increases in the interval 7 – 20 units of polyphosphate monomer. Save for this critical value, the degree of polymerization has an insignificant effect on the metachromatic interaction [36-39].

2.3.2. The Structural Parameters of the Colorant

2.3.2.1. The Influence of the Stearic Factors

The shift of the spectral bands of the conjugated chromophores, compared with the isolated ones, is possible only within an unsaturated plane system. The noncoplanarity of a conjugated system induced by the specific geometry of a molecule has, as an effect, the shift towards an independent absorption of light, characteristic to isolated chromophores. In this situation we observe a hypsochrome shift of the absorption maximums. The charge-transfer complexes can be identified by means of UV-Vis spectra on account of the characteristic bands appearing alongside the bands corresponding to neutral molecules [25, 35, 40].

The appearance of charge-transfer bands can be caused by the transitions from the fundamental state, in which the charge is practically undistributed between the two components, to an excited state, the predominant structure of which is the one corresponding to the complete transfer of an electron from a donor molecule to an acceptor molecule [41].

2.3.2.2. The Absorption Characteristics of Metachromatic Colorants. The Effects of the Electronic Structure

In the case of benzenic chromophores, the bathochromic shifts are attributed to the hyperconjugation between the σ electrons of an alkyl group found in resonance with the benzenic ring [2, 35, 40, 42].

The methyl groups are efficient in hyperconjugation. The presence of a double, conjugated bond, like in styrene, leads to the appearance of an intense band, shifted bathochromically. The shift and intensity of the absorption band is due to the intervention of the mezomere structures, with separate charges, in the excited state of the molecule, which accentuates as the conjugation increases.

The auxochrome groups with electron-repelling (-E) effect, such as -NH_2, -NR_2, -OH, -X (halogen), produce a bathochromic shift of the absorption band. This shift can be caused by the conjugation of nonparticipatory electrons of the auxochrome group with the π electrons of the aromatic core. On account of this conjugation, the electronic density at the level of the aromatic core increases, which diminishes the energy required for the electronic transition. Accordingly, the absorption band shifts bathochromically [25, 35, 40].

The blocking of the nonparticipatory electrons, for instance through protonation, cancels the bathochromic effect. In the case of the auxochrome, the electronic density from the aromatic core depends on the +E effect. The bathochromic shift increases as the electronegativity of the atom with nonparticipatory electrons decreases.

The auxochrome groups of the type -NH$_2$, -NR$_2$, -OH, -OR, etc. possess nonparticipatory electrons capable of interacting with the π electrons of the aromatic chromophore systems.

The electron-donor substituents *cause the appearance* of a *bathochrome effect by conjugating the nonparticipatory electrons with the π-aromatic electrons*. When the electrical density at the core increases, the electronic transition can take place with smaller energies; accordingly, absorption appears at greater wavelengths. The *bathochrome shift* is *stronger* as the *heteroatom of the auxochrome produces in the core a greater density of electrons*. By substituting with electron-donor groups, simultaneously with the bathochromic shift, we also observe a decrease in absorption [43-45].

The shift of the spectral bands of the conjugated chromophores is possible only in the conditions of a plane unsaturated system. In general, the coplanarity of a conjugated system caused by the specific geometry of a molecule induces a hypsochromic shift of the maximums of the absorption bands [25, 40].

The unexpected hypsochrome effect in the case of accumulation of auxochrome groups registered at the absorption of the CV light is due to the symmetry of the molecule that provides the common electronic system around the central carbon atom. Accordingly, some quantic energetic states are identical, their number decreases, and the differences between the state of fundamental energy and the excited energetic states are reflected in the absorption of light at smaller wavelengths, thus producing the hypsochrome effect observed spectrally [46]. This effect has also been confirmed by our recent studies that present the result of the interaction between, on the one hand, the copolymer of maleic acid with vinyl acetate in the form of sodium salt (NaM-AV) [47, 48], and, respectively, the copolymer of maleic acid with styrene in the form of sodium salt (NaM-S) [49-51], as anionic partners, and Crystal Violet (triphenylmethanic colorant) using UV-Vis spectroscopy.

This is not the case with the colorant Victoria Blue B (triarylmethanic colorant), the molecule of which is asymmetric, a consequence of the bathochrome effect also observed in other recent studies [52-54].

Metachromasia is a complicated phenomenon involving multiple bonds between the colorant and the polyanion [55, 56]. In this situation, the change in the absorption bands in the spectrum of the metachromatic colorant in the presence of polyelectrolytes *is specific to each polyelectrolyte*. It has also been shown that the configuration of the π electrons of the bonded colorant can influence foremost the distance between the bonded positions of a polyanion [56, 57-58].

2.3.3. Influence of pH on the Metachromatic Phenomenon

The UV-Vis spectra of the metachromatic complexes are influenced by the concentration of hydrogen ions, changing their appearance as the pH changes. These changes are frequently found in the case of compounds in which the chromophore or auxochrome groups dissociate or enolize (-COOH, $-NH_2$, -OH phenolic, etc.). The change in pH in the case of these compounds induces the change of the position and intensity of the bands, or leads to the appearance of new bands [25, 41].

For the quantitative study of the metachromatic phenomenon in the case of the Crystal Violet–sodium poly(α-L-glutamic) and/or poly(α-L-glutamate) by means of the PCA technique, we resorted to unfolding the spectra in order to determine the relative quantities of CV in the three different media: in solution, α-helicoidal chaining (α–helix chaining) at a low pH or greater values of the pH. The results show that the distinction of the colorant's spectrum can be achieved in all three media [3].

In the case of pseudocyanine and pinacol in the presence of sodium poly-L-glutamate, the metachromatic phenomenon is dependent on the pH. Thus, the metachromatic shifts are distinct and pronounced in basic solutions, with the polymer molecules being strongly charged, relaxed and elongated. In acidic solutions the metachromatic shifts are weak, and the acidic groups on the polymer are protonated [3].

2.3.4. Influence of Solvents on the Metachromatic Phenomenon

Multiple studies have shown that the colorant–colorant interactions are responsible for the metachromatic phenomenon, and that the interactions between the water molecules play a key role in this phenomenon [59, 60].

The spectral changes observed at dehydration can be explained thus: the cations of the colorant come very close to the anionic positions and as a result the colorant–colorant interactions are drastically reduced and become depended on the colorant–colorant *distance*. The anionic celluloses produce hypsochromic shifts in the colorant's absorption spectrum as follows:

sulfate > phosphate > carboxyl > carboxymethyl

Thus, by dehydrating a colorant–polyphosphate complex, the metachromasy disappears, since the cations of the colorant and the free charged positions on the polyanion lose their hydration water and thus the colorant–anionic position increases.

The colorant–colorant interaction is so much reduced that its absorption spectrum resembles that of an isolated set of cations of the colorant. This hypothesis explains the fact that *some colorants are not metachromatic*.

Urea and ethanol destroy the hydrophobic bonds, but compared with ethanol it increases the water's dielectric constant [59]. Most often, for metachromasy to be efficiently destroyed, the electrolyte must be several hundred times more concentrated than the colorant.

Solvents with moderate dielectric constant such as ethanol and acetone induce a metachromatic behavior of the colorants only when found at high concentrations. Cetylpyridinium chloride destroys the metachromasy of the colorants Methylene blue and Crystal violet, induced by the chondroitin sulfate and heparin. Thus, cetylpyridinium chloride replaces the colorant ions on the polyanion, at an almost stoichiometric ratio [61].

2.3.5. *Influence of Salt on the Metachromatic Phenomenon*

The metachromatic phenomenon induced by adding salt is similar to that occurring when the concentration of the colorant increases. The effect of adding salt (NaCl) in an aqueous solution of Methylene blue is observed in the absorption spectrum by the fact that the aggregation of the colorant is dependent on the concentration of salt. Many colorants behave like

Methylene blue in the presence of salts, but in some cases there occur precipitations or the appearance of formations resembling colloidal structures. Such phenomena clearly show that the H bands can appear at the interaction between the colorant and the counterion. This is also supported by the fact that $BaCl_2$ and $CaCl_2$ show double efficiency than NaCl, and the sulfates of divalent metals are more efficient than their respective chlorides [62-64].

2.3.6. Influence of Reaction Time, Temperature and Pressure

In many studies on the metachromatic phenomenon, the metachromatic system is agitated from several minutes up to an hour, in order to stabilize the reaction [1, 3].

The mechanism behind the destruction of the metachromatic phenomenon depends on temperature: as the temperature of the metachromatic system increases, the colorant can be dislodged from the polyanion by breaking the electrostatic bonds between the colorant cations and the polyanion; by lowering the temperature of the metachromatic system, a characteristic sharp metachromatic band appears. Most often, the work is performed at room temperature (25°C) [2].

The effect of pressure on the association of the Rhodamine B and Methylene blue molecules has been studied at room temperature and a hydrostatic pressure of 4500 atm. It was found that as the hydrostatic pressure increases, so does the association of the two colorants. This association of the colorant molecules is also influenced by the increase in the intensity of the London dispersion forces, of the water's dielectric constant simultaneously with the increase in pressure [22].

3. DYEING MELANA FIBERS WITH CV IN THE PRESENCE OF MALEIC ANIONIC POLYELECTROLYTES

Studies on the process of dyeing textile materials with anionic colorants using cationic retarders are numerous, but in the case of dyeing textile materials with cationic colorants that use as anionic retarders compounds based on polyelectrolytes, the specialized literature is much more scarce [65].

In this sense, studies have been carried out on dyeing Melana fibers with cationic colorants of the type Crystal violet (CV) in the presence of retarders based on the copolymer of maleic acid with vinyl acetate (NaM-AV) or the copolymer of maleic acid with styrene (NaM-S) [66].

3.1. Materials Used

3.1.1. Romanian Melana Fiber

Romanian Melana fiber is a polyacrylonitrilic fiber based on a ternary polymer (85% acrylonitrile, 10% vinyl acetate, 5% α-methylstyrene), obtained through a radical polymerization reaction initiated in a potassium persulfate–sodium metabisulfite redox system [67].

Schematically, the structure of the Melana fiber can be represented as in Figure 5.

The main properties are listed in Table 1 [68, 69].

3.1.2. Polyelectrolytes

3.1.2.1. The Polyelectrolyte (NaM-AV)

The polyelectrolyte (NaM-AV) is a copolymer of maleic acid with vinyl acetate in the form of sodium salt [47, 70].

The chemical structure of the sodium maleate–vinyl acetate copolymer is presented in Figure 6.

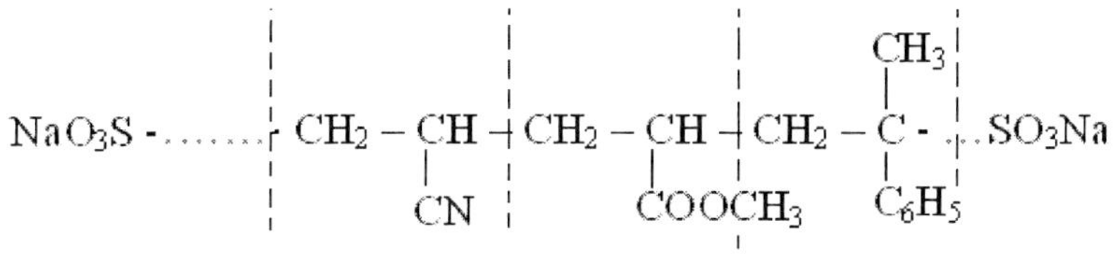

$$NaO_3S \cdots CH_2-CH + CH_2-CH + CH_2-C(CH_3)(C_6H_5) \cdots SO_3Na$$
$$(CN)\quad (COOCH_3)$$

Figure 5. The schematic structure of the Melana fiber.

Table 1. Properties of the Melana fiber

No.	Characteristic	UoM	Value
1.	Density at 21°C	g/cm^3	1.16
2.	Reprise	%	2
3.	Tenacity	g/den	2.5 – 3.0
4.	Elongation at break	%	40 – 55
5.	Fiber wrinkling	wrinkles/cm	3 – 4
6.	Level of whiteness (on on the leucometer with blue filter)	%	70 – 73
7.	Ondulations	No./cm	3 – 4
8.	Contraction at 100°C	%	18 – 22
9.	Resistance to microorganisms and insects		very good
10	E_A acidic equivalent	[µE/g]	47.01
11	E_B basic equivalent	[µE/g]	9.24
12	S_{rel} relative saturation		1.88

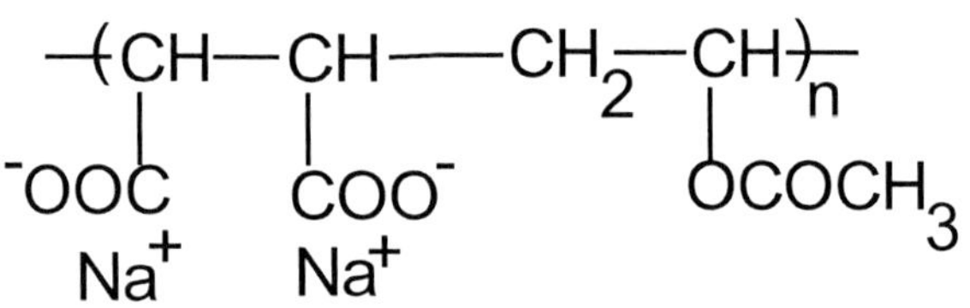

$$-(CH-CH-CH_2-CH)_n-$$
$$|||$$
$$^-OOCCOO^-OCOCH_3$$
$$Na^+Na^+$$

Figure 6. The chemical structure of the sodium maleate–vinyl acetate copolymer (NaM-AV).

Average molecular mass Mv = 70000

Recent studies have shown that the copolymer of maleic acid with vinyl acetate has been used with notable results in [71-73]:

- bone implants, by improving the characteristics of the bioadhesive surface;
- soil stabilization, by forming fine films on the surface of the structural elements, by forming bridges between soil aggregates;
- inhibiting the deposition of carbonate crusts from geothermal waters.

3.1.2.2. The Polyelectrolyte (NaM-S)

The polyelectrolyte (NaM-S) is a copolymer of maleic acid with styrene, in the form of sodium salt [74].

The chemical structure of the sodium maleate–vinyl acetate copolymer is presented in Figure 7.

Average molecular mass Mv = 95000

Relatively recent studies have shown the successful use of the maleic acid–styrene copolymer in stabilizing soil structure by increasing hydrostability [72].

The polyelectrolyte (NaM-S) was used as a sodium salt obtained in the laboratory according to a method described in the literature [74].

3.1.3. The Crystal Violet (CV) Colorant

The Crystal violet (CV) colorant, Basic Violet 3, C. I. 42555 [75].

The chemical structure of the colorant Crystal violet is presented in Figure 8:

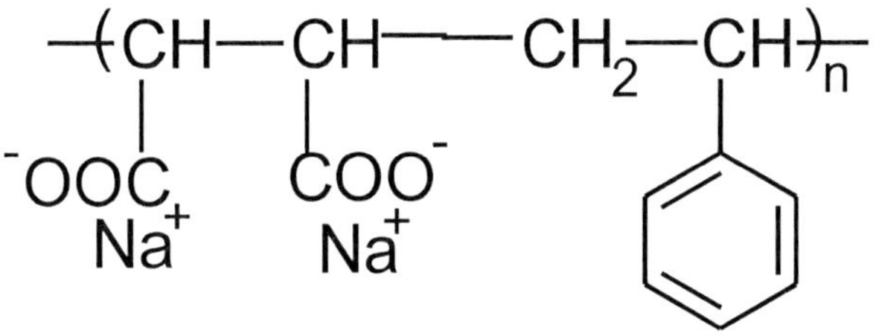

Figure 7. The chemical structure of the sodium maleate–styrene copolymer (NaM-S).

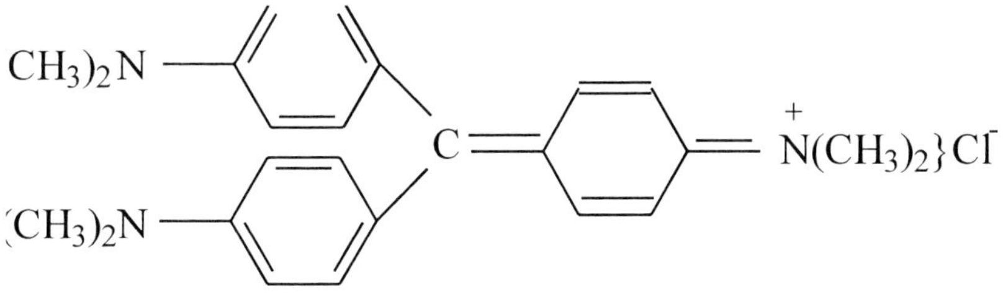

Figure 8. The chemical structure of the colorant Crystal Violet.

Class of colorants to which it belongs: triaminotriphenylmethanic colorants

Chemical formula: $C_{25}H_{30}N_3Cl$

Molecular mass: 408 g/mol

Chemical designation: N-[4-[bis[4-(dimethylamino)-phenyl]methylene]-2,5-cyclohexadiene-1-ylidene]-N-methyl-methenamine chloride, or Hexamethyl p-rosaniline chloride

Solubility: soluble in water, alcohol, glycerin, insoluble in ether.

Uses: Crystal violet is used for the following [46, 76]:

- colorizing inks and tusches;
- dyeing paper, silk, food;
- in cytology;
- in animal and plant histology;
- in mycology;
- as a bacteriostatic agent.

3.2. Particularities of Melana Fiber Dyeing

The technological process for dyeing Melana fiber requires that certain conditions are met [77, 78]: the appropriate recipe with respect to the saturation value; selecting the colorants that have identical or close K indexes; dissolving the colorants in the conditions prescribed by the colorant cards; the strict observance of the temperature and time regimen; compliance with the drying temperature at maximum 60°C; the mass of the Melana sample equal and uniform.

3.2.1. The Control of the Dyeing Speed by Adding Equalization and Slowing Agents

The affinity of the cationic colorants for acrylic fibers is very large on account of the electrostatic attractions as well as of the hydrophobic interactions between the fiber and the colorant, which constitutes a great difficulty for obtaining uniform dyeing on acrylic fibers [69, 79-91].

Very important is the precise control of the temperature over the T_{II} transition temperature [91]. In the 90 – 98°C interval, any increase by 1°C of the temperature causes an increase of the adsorption speed by 30%; this is due to the fact that an increase in temperature causes a large increase

Aspects Concerning the Hypsochrome Effect ... 21

(according to the Arrhenius equation) in the activation energy [80, 83]. In order to obtain uniform dyeing, the speed of temperature increase must be very carefully controlled after reaching the temperature of vitrification. When this is not possible, equalization and slowing agents.

The slowing agents or retarders can be: cationic, anionic, polymeric [94].

3.2.2. Anionic Retarders

Anionic retarders act on the colorants by forming a complex with a slower diffusion in solution speed, thus a more sluggish absorption. These products do not have selective action, but diminish the tinctorial efficiency by also retaining the colorant in the solution in extreme cases by precipitating it [95].

The mechanism for bonding the colorant can be represented schematically as follows [96]:

$$Col^+ + Ret^- \underset{>80°C}{\overset{<65°C}{\rightleftarrows}} Col - Ret$$

$$Col^+ + PAN^- \underset{>95°C}{\overset{\geq 65°C}{\rightleftarrows}} Col - PAN$$

where Col^+ – dyes cation; Ret^- – retarder (polyelectrolyte) anion; PAN^- – free anionic groups from the ends of polyacrylonitrile macromolecular chains.

Such data concerning the dyeing of Melana fibers in aqueous solutions (distilled water), with CV on a water bath, in the presence of retarders NaM-AV or NaM-S were presented in recent lectures and some are presented in Table 2 [95, 97].

The optimal correlations between the application conditions and the color differences of the Melana fiber for industrial appliance of anionic retarders based on maleic copolymers using the mathematical modelling elements, have been presented recently [98].

Table 2. Remissions corresponding to the Melana samples dyed with CV in the presence of the two NaM-S and NaM-AV polyelectrolytes in the domain λ = 400 ÷ 700 nm

No.	λ,nm	std CV	0.5% NaM-S	1.5% NaM-S	2.5% NaM-S	0.5% NaM-AV	1.5% NaM-AV	2.5% NaM-AV
1	400	**77.89**	96.42	93.49	90.26	84.75	84.16	83.53
2	410	**71.43**	103.38	100.17	96.27	77.83	77.1	76.94
3	420	**67.12**	104.81	102.06	97.12	71.56	71.13	71.01
4	440	**64.61**	92.99	93.91	85.16	62.54	64.0	62.92
5	450	**63.42**	81.94	84.76	74.66	57.13	59.76	57.64
6	470	**55.60**	54.33	60.07	49.05	41.77	46.07	42.18
7	480	**47.9**	39.26	45.38	35.63	32.2	36.75	32.47
8	500	**32.12**	18.64	23.32	17.16	17.28	20.94	17.22
9	510	**24.76**	12.27	15.93	11.38	11.7	14.73	11.65
10	530	**15.32**	6.17	8.32	5.73	5.96	7.86	5.81
11	540	**12.88**	5.05	6.77	4.66	4.75	6.36	4.67
12	560	**11.03**	4.50	5.90	4.13	4.02	5.38	3.95
13	570	**10.91**	4.59	5.98	4.20	3.98	5.31	3.89
14	590	**10.44**	4.83	6.15	4.24	3.86	5.15	3.84
15	600	**10.4**	4.98	6.26	4.37	3.82	5.16	3.82
16	620	**14.5**	7.78	9.85	6.81	5.66	7.72	5.81
17	630	**20.14**	12.20	15.29	10.87	8.76	11.79	9.19
18	620	**14.5**	7.78	9.85	6.81	5.66	7.72	5.81
19	630	**20.14**	12.20	15.29	10.87	8.76	11.79	9.19
20	650	**39.52**	32.50	38.29	29.44	23.81	29.62	25.3
21	660	**52.60**	48.54	54.86	44.39	36.82	43.48	38.89
22	680	**80.21**	86.33	91.35	81.52	71.04	76.13	73.06
23	690	**90.35**	102.20	105.62	97.90	86.8	89.62	87.94
24	700	**97.85**	114.03	115.47	110.30	99.67	99.98	99.55

Considering the fact that triaminotriphenylmethanic colorants (for example, Crystal violet) present very pronounced remission phenomena so that the color of their crystals (greenish) is complementary to the color of the solutions they form (red-violet) [99], we analyzed the effect produced by the two polyelectrolytes (NaM-S, NaM-AV) on the Melana fibers dyed with CV. The results are presented in Table 2.

Table 3. K/S values corresponding to the samples of Melana dyed with CV in the presence of NaM-S sau NaM-AV at λ = 400 ÷ 450 nm

No.	λ, nm	R% Std CV	K/S std	R% 0.5% NaM-S	K/S 0.5% NaM-S	R% 1.5% NaM-S	K/S 1.5% NaM-S	R% 2.5% NaM-S	K/S 2.5% NaM-S	R% 0.5% NaM-AV	K/S 0.5% NaM-AV	R%, 1.5% NaM-AV	K/S 1.5% NaM-AV	R% 2.5% NaM-AV	K/S 2.5% NaM-AV
1	400	77.89	0.0314	96.42	0.00067	93.49	0.00226	90.26	0.00527	84.75	0.0137	84.16	0.0149	83.53	0.0163
2	410	71.43	0.0571	103.38	0	100.17	0	96.27	0.00073	77.83	0.0316	77.1	0.034	76.94	0.0346
3	420	67.12	0.0806	104.81	0	102.06	0	97.12	0.00043	71.56	0.0565	71.13	0.0586	71.01	0.0592
4	430	65.99	0.0877	101.14	0	99.92	0	93.07	0.00259	67.7	0.0771	68.02	0.0752	67.64	0.0774
5	440	64.61	0.0970	92.99	0.00263	93.91	0.00198	85.16	0.0129	62.54	0.1123	64.0	0.1013	62.92	0.1181
6	450	63.42	0.1055	81.94	0.002	84.76	0.0137	74.66	0.0321	57.13	0.1609	59.76	0.1353	57.64	0.1559

After analyzing the data from the table above, we can state that in the case of the Melana samples dyed with CV in the presence of NaM-S or NaM-Av, at small wavelengths remission is due to the differences in the hydrophilic-hydrophobic character of the two polyelectrolytes, to the steric hindrance caused both by the remnant styrene from the composition of the NaM-S polyelectrolyte, as well as to the fact that, in general, triaminotriphenylmethanic colorants (e.g., Crystal violet) present very pronounced remission phenomena [96].

For establishing if NaM-S or NaM-AV present retarding effect on Melana fibers dyed with CV, the values of the K/S ratios [96-98] were calculated, corresponding to the remissions from the 400 ÷ 450 nm domain in which CV presents maximum absorption. The results are presented in Table 3. We observe that at λ = 400 nm (wavelength at which the colorant CV presents maximum absorption), all the K/S values corresponding to the Melana samples dyed with CV in the presence of NaM-S or NaM-AV are smaller than those of the standard-dyed Melana samples.

The NaM-S and NaM-AV polyelectrolytes present retarding effect irrespective of the concentration at which they are found in processes of dyeing Melana with CV, on account of the voluminosity of the colorant molecules as well as of the hydrophoby of these systems. The phenomenon of steric hindrance is more pronounced in the case of the Melana samples dyed in the presence of NaM-S because of the styrene radical. Moreover, the K/S values are very small because of the very large remissions displayed by these samples of Melana dyed with CV [100].

CONCLUSION

The main factors influencing the metachromatic phenomenon have been presented.

The shift of the spectral bands of the conjugated chromophores is possible only in the conditions of a plane unsaturated system, such as Crystal Violet. In general, the coplanarity of a conjugated system induced by the specific geometry of a molecule induces a hypsochromic shift of the absorption bands' maximums.

In the case of the Melana samples dyed with CV, the presence of NaM-S or NaM-AV at small wavelengths is due to the difference in the hydrophile-hydrophobe character of the two polyelectrolytes, of the stearic hindrance caused by both the styrene remnants from the composition of the NaM-S polyelectrolyte, and the fact that the triaminophenylmethanic colorants (e.g., Crystal violet) have very pronounced remission phenomena.

The NaM-S and NaM-AV polyelectrolytes manifest retardation effect irrespective of the concentration at which they are found in processes of dyeing Melana with CV, on account of the voluminosity of the colorant molecules and the hydrophoby of these systems.

REFERENCES

[1] Kugel, R. *Focus on Ionic Polymer, 3, Metachromasy as an indicator of polycation conformations in aqueous solution*, Ed. E.S. Drăgan, pg. 93, 2005.

[2] Bercu, E. PhD Thesis, *Gheorghe Asachi Technical University Iasi*, pg. 5-17.

[3] Kugel, RW. Advances in Chemistry Series, 236, *Structure-Property Relations in Polymers*, pg. 507, 1993.

[4] Masci, G; Barbetta, A; Dentini, M; Crescenzi, V. *Macromol Chem. Phys.*, 200, pg.1157, 1999.

[5] Peyratout, C; Donath, E; Daehne, L; Photochem, J. *Photobiol A: Chem.*, 142, pg. 51, 2001.

[6] Georges, J. *Spectrochim. Acta*, 51 A, pg. 985, 1995.

[7] Nandini, R; Vishalakshi, B. Spectrochim. *Acta Part A: Molecular and Biomolecular Spectroscopy*, 74(5), pg. 1025, 2009.

[8] Antonov, L; Gergov, G; Petrov, V; Kubista, M; Nygren, J. *Talanta*, 49, pg. 99, 1999.

[9] Luek, HB; Rice, BL; McHale, JL. *Spectrochimica Acta*, 48 A(6), pg. 818, 1992.
[10] Coon, SR; Zakharian, TY; Littlefield, NL; Loheide, SP; Puchkova, EJ; Freeney, RM; Pak, VN. *Langmuir*, 16, pg. 9690-9693, 2000.
[11] Luek, HB; Rice, BL; McHale, JL. *Spectrochimica Acta*, 48A(6), pg. 818, 1992.
[12] Neumann, MG; Hioka, N. *J. of Appl., Polym. Sci.*, 34(8), pg. 2829, 1987.
[13] Park, CH; Park, MS; Lee, H. *Taehan Hwahakhoe Chi*, 31(4), pg. 295, 1987.
[14] Homem-de-Mello, P; Mennucci, B; Tomasi, J. *Theor. Chem. Acc*, 118, pg. 305, 2007.
[15] Zahrim, AY; Tizaoui, C; Hilal, N. *Desalination*, 266, pg. 1, 2011.
[16] Shirai, M; Nagaoka, Y; Tanaka, M. *J. Polym. Sci., Polym. Chem.*, Ed 15, pg. 1021, 1977.
[17] Saletskii, AM; Tkachev, AM. *Optics and Spectroscopy*, 93(2), pg. 232, 2002.
[18] Yamaoka, K; Takatsuki, M; Nakata, K. *Bull. Chem. Soc. Jpn.*, 53, pg. 3165, 1980.
[19] Shirai, M; Nagatsuka, T; Tanaka, M. *Makromol. Chem.*, 178, pg. 37, 1977.
[20] Bercu, E. PhD Thesis, *Gheorghe Asachi Technical University Iasi*, pg. 13, 2011.
[21] Pal, MK; Schubert, M. *J. Phys. Chem.*, 67, pg. 1821, 1963.
[22] Pal, MK; Bhattacharyya, AK. *Makromol Chem.*, 185, pg. 2241, 1984.
[23] Luo, HQ; Liu, SP; Liu, ZF; Liu, Q; Li, NB. *Anal Chem. Acta*, 261, pg. 449, 2001.
[24] Yamaoka, K; Takatsuki, M. *Bull. Chem. Soc. Jpn.*, 51(11), pg. 3182, 1978.
[25] Silverstein, BM. *Spectrometric identification of organic compounds*, Fifth Ed., John Wiley &Sons, Inc., New York., pg. 289, 1991.
[26] Quadrifoglio, F; Crescenzi, V. *J. Col. Int. Sci.*, 35(3), 447, 1971.
[27] Pal, MK; Chaudhuri, M. *Makromol. Chem.*, 133, pg. 151, 1970.

[28] Bercu, E. PhD Thesis, *Gheorghe Asachi Technical University Iasi*, pg. 14, 2011.
[29] Takatsuki, M. *Bull. Chem. Soc. Jpn.*, 53, 1922, 1980.
[30] Shirai, M; Yamashita, M; Tanaka, M. *Makromol. Chem.*, 179, pg. 747, 1978.
[31] Mueller, M; Meier-Haack, J; Schwarz, S; Buchhammer, HM; Eichhorn, KJ; Janke, A; Kessler, B; Nagel, J; Oelmann, M; Reihs, T; Lunkwitz, K. *J. of Adhesion*, 80(6), pg. 521, 2004.
[32] Gummow, BD; Roberts, GAF. *Makromol. Chem.*, 187, pg. 995, 1986.
[33] Yamaoka, K; Hashimoto, H. *Chem. Lett.*, pg. 465, 1976.
[34] Lin, X; Zhong, A; Chen, D; Zhou, Z; He, B. *J. Appl. Polym. Sci.*, 87, pg. 369, 2003.
[35] Scutaru, D. *Metode spectrale utilizate în analiza structurală organică* [*Spectral methods used in organic structural analysis*], Rotaprint, Iasi, pg. 89, 1994.
[36] Wainwright, M; Burrow, SM; Phoenix, AD. *J. Waring, Cytotechnology*, 29, pg. 35, 1999.
[37] Demirbas, O; Alkan, M; Dogan, M. *Adsorbtion*, 8, pg. 341, 2002.
[38] Ruprech, J; Baumgaertel, H. *Ber. Bunsenges. Phys. Chem.*, 88, pg. 145, 1984.
[39] Hochberg, GC. *Colloid and Polym Sci.*, 272, pg. 409, 1994.
[40] Balaban, AT; Banciu, M; Pogany, I. *Aplicaţii ale metodelor fizice în chimia organică* [*Applications of physical methods in organic chemistry*]. Ed. Ştiinţifică şi Enciclopedică, Buc., pg. 39, 1983.
[41] Dubois, F. Boue, *Macromolecules*, 34, pg. 3684, 2001.
[42] Ortona, O; Vitagliano, V; Sartorio, R; Costantino, L. *J. Phys. Chem.*, 88, pg. 3244, 1984.
[43] Cîdu, E; Chiţanu, GC; Grigoriu, A; Anghelescu-Dogaru, AG. *Bull. UTI, Tomul L(LIV), Fasc.*, 1-2, pg. 81, 2004.
[44] Kugel, RW. Advances in Chemistry Series, 236, *Structure-Property Relations in Polymers*, pg. 507-533, 1993.
[45] Bercu, E. PhD Thesis, *Gheorghe Asachi Technical University Iasi*, pg. 41, 2011.

[46] Floru, L; Urseanu, F; Tărăbăşanu, C; Palea, R. *Chimia și tehnologia intermediarilor aromatici și a coloranților organici*, Ed. Didactică și Pedagogică, Buc., pg 432, 1980. [*Chemistry and Technology of Aromatic Intermediates and Organic Dyes*]

[47] Chitanu, CG; Carpov, A; Asaftei, T. *Brevet Rom. Nr.*, 106745, 1993.

[48] Chitanu, GC. *Studiul proprietatilor fizico-chimice ale copolimerilor anhidridei maleice*, Teza de doctorat, Iasi, pg. 70, 1995. [*Study of physicochemical properties of maleic anhydride copolymers*]

[49] Cîdu, E; Chiţanu, GC; Grigoriu, A. *Bull. UTI, Tomul LI (LV), Fasc.*, 1-2, pg. 91, 2005.

[50] Cîdu, E; Radu, CD; Chiţanu, GC; Grigoriu, A. *The 6th International Conference Management of Technological Changes Alexandropolis-Grecia*, pg. 49, 2009.

[51] Cîdu, E; Chitanu, GC; Grigoriu, A. *Acta Universitatis Cibiniensis, Seria F Chemia*, 8(1), pg. 5, 2005.

[52] Bercu, E; Sandu, I; Aldea, HA; Vasilache, V; Toma, V. *Rev. Chem.*, 64(10), pg. 1121, 2013.

[53] Bercu, E. PhD Thesis, *Gheorghe Asachi Technical University Iasi*, pg. 51-74, 2011.

[54] Cîdu, E; Radu, CD; Chiţanu, GC; Grigoriu, A. *The 6th International Conference Management of Technological Changes Alexandropolis-Grecia*, pg. 45, 2009.

[55] Kugel, R. *Focus on Ionic Polymer*, pg. 93, 2005.

[56] Zhang, L; Li, N; Zhao, F; Li, K. *Analytical Sciences*, 20, pg. 445, 2004.

[57] Wainwright, M; Burrow, SM; Guinot, SGR; Waring, J. *Cytotechnology*, 29, pg. 35, 1999.

[58] Pal, MK; Ghosh, BK. *Makromol. Chem.*, 192, pg. 467, 1991.

[59] Tsun, Sin; Yui, TI; Pal'mer, VG; Musabekov, KB. *Kolloidnyi Zhurnal*, 49(4), pg. 819, 1987.

[60] Hayakawa, K; Ohta, J; Maeda, T; Satake, I. Jan Kwak, CT. *Langmuir*, 3(3), pg. 377, 1987.

[61] Pal, MK; Ghosh, BK. *Makromol. Chem.*, 192, pg. 467, 1991.

[62] Pal, MK; Mandal, N. *Macromol. Chem.*, 190, pg. 2501, 1989.

[63] Pal, MK; Basu, S. *Makromol. Chem.*, 27, pg. 69, 1958.
[64] El-Sayed, GO. *Desalination*, 272 (1-3), pg. 225, 2011.
[65] Voncina, B; Vivod, V; Jausovec, D. *Dyes and Pigments*, 74, pag. 642-646, (2007).
[66] Bercu, E. PhD Thesis, *Gheorghe Asachi Technical University Iasi*, pg. 76-138, 2011.
[67] Grigoriu, A; Coman, D. *Bazele finisării materialelor textile*, Ed. Tehnopress, Iaşi, 2001. [*The basics of textile finishing*]
[68] Grigoriu, A; Stoichiţescu, L. *Tehnologia Chimică Textilă*, Rotaprint, Iaşi, 1992. [*Textile Chemical Technology*]
[69] Popescu, V. *Cercetări privind îmbunătăţirea calităţii fibrelor PAN, finisate, destinate industriei tricotajelor*, Teză de doctorat Iaşi, pg. 53, 1998. [*Researches on improving the quality of PAN, finished, for the knitwear industry*]
[70] Chitanu, CG. *Studiul proprietatilor fizico-chimice ale copolimerilor anhidridei maleice*, Teza de doctorat, Iasi, pg. 70-73, 1995. [*Study of physicochemical properties of maleic anhydride copolymers*]
[71] Sima, LE; Filimon, A; Piticescu, RM; Chitanu, GC; Suflet, DM; Miroiu, M; Socol, G; Mihailescu, IN; Neamtu, J; Negroiu, G. *J. Mater. Sci.: Mater Med.*, 20, pg. 2305, 2009.
[72] Piticescu, M; Popescu, LM; Giurgincă, M; Chitanu, GC; Negroiu, G. *Journal. of Optoelectronics and Advanced Materials*, 9(11), pg. 3340, 2007.
[73] Negroiu, G; Piticescu, RM; Chitanu, GC; Mihailescu, IN; Zdrenţu, L; Miroiu, M. *J. Mater. Sci.: Mater Med.*, 19, pg. 1537, 2008.
[74] Houben-Weyl, "*Methoden der Organischen Chemie*" [*Methods of Organic Chemistry*], E 20, "*Macromolekulare Stoffe*" [*Macromolecular substances*], Georg Thieme Verlag, Stuttgart, pg. 1239, 1987.
[75] *Colour Index*, 4, Ed. 3, pg. 4391, 1971.
[76] Eldem, Y; Omer, I. *Dyes and Pigments*, 60, pg. 49, 2004.
[77] Kim, YH; Sun, G. *Tex. Res. J.*, 72(12), pag. 1052, 2002.
[78] Bercu, E. PhD Thesis, *Gheorghe Asachi Technical University Iasi*, pg. 41, 2011.

[79] Yang, Y. *J. Soc. Dyers. Col.*, 110(3), pg. 98, 1994.
[80] AitKen ş.a, D. *J. Soc. Dyers. Col.*, 108(4), pg. 219, 1992.
[81] Kim ş.a, JP. *J. Soc. Dyers. Col.*, 111(4), pg. 107, 1995.
[82] Dawson, TL. *J. Soc. Dyers. Col.*, 97(3), pg. 115, 1981.
[83] Shukla ş.a, SR. *J. Soc. Dyers. Col.*, 107(11), pg. 407, 1991.
[84] Shukla ş.a, SR. *J. Soc. Dyers. Col.*, 107(12), pg. 463, 1991.
[85] Shukla ş.a, SR. *J. Soc. Dyers. Col.*, 108(1), pg. 29, 1992.
[86] Achi ş.a, SS. *J. Soc. Dyers. Col.*, 111(10), pg. 328, 1995.
[87] Gur-Arieh, Z. *J. Soc. Dyers. Col.*, 90(1), pg. 8, 1975
[88] Gur-Arieh, Z. *J. Soc. Dyers. Col.*, 92(9), pg. 332, 1976.
[89] Gur-Arieh, Z. *J. Soc. Dyers. Col.*, 90(1), pg. 12, 1974.
[90] Kim ş.a, JP. *J. Soc. Dyers. Col.*, 111(4), pg. 107, 1995.
[91] Bajaj ş.a, P. *Tex. Res. J.*, 60(2), pg. 113, 1990.
[92] Volko, EI. *Review of Progress in Coloration*, 20, pg. 50, 1990.
[93] Shukla ş.a, SR. *J. Soc. Dyers. Col.*, 109(10), pg. 330, 1993.
[94] Bercu, E. PhD Thesis, Gheorghe Asachi Technical University Iasi, pg. 40, 2011.
[95] Stoichiţescu, LI. *Bazele teoretice şi practice ale vopsirii şi imprimării materialelor textile*, Ed. Performantica, Iaşi, pg. 42, 2002. [*Theoretical and practical basics of textile printing and printing*]
[96] Bercu, E; Diaconescu, RM; Radu, CD; Popescu, V. *European J. of Sci. and Theology*, 8, pg. 235-238, 2012.
[97] Bercu, E. PhD Thesis, *Gheorghe Asachi Technical University Iasi*, pg. 76, 2011.
[98] Bercu, E. PhD Thesis, *Gheorghe Asachi Technical University Iasi*, pg. 78-85, 2011.
[99] Guneri, P; Epstein, JB; Ergun, S; Bozacioglu, H. *Clin. Oral. Invest.*, 15, pg. 337, 2011.
[100] Bercu, E. PhD Thesis, *Gheorghe Asachi Technical University Iasi*, pg. 150-151, 2011.

In: Crystal Violet
Editor: Victor Duffet
ISBN: 978-1-53615-806-9
© 2019 Nova Science Publishers, Inc.

Chapter 2

ENVIRONMENTALLY SAFE BIOSORBENTS FOR CRYSTAL VIOLET REMOVAL FROM WASTEWATER

Preeti Pal[1,*], *Asmaa Benettayeb*[2,3,†] *and Anjali Pal*[1,4,‡]
[1]School of Environmental Science and Engineering,
Indian Institute of Technology Kharagpur, India
[2] Laboratoire de Génie Chimique et de catalyse hétérogène,
Université de Sciences et de la Technologie -Mohamed Boudiaf,
USTO-MB, Oran, Algérie
[3]Ecole des Mines d'Alès, Centre des Matériaux des Mines d'Alès
(C2MA), France
[4]Civil Engineering Department, Indian Institute of Technology,
Kharagpur, India

ABSTRACT

Crystal violet (CV) is a triarylmethane dye; widely applied in coloring paper, as a temporary hair colorant, for dyeing cotton, silk, and

---

[*] E-mail: pal.preiti@iitkgp.ac.in.
[†] E-mail: asma.benettayeb@gmail.com.
[‡] Corresponding Author's E-mail: anjalipal@civil.iitkgp.ac.in.

wool. It is a cationic dye which is also known as gentian violet. CV is used for the manufacture of paints and printing inks; it is also used as a biological staining agent. In animal and veterinary medicine, it is employed as an antibacterial, antifungal, and antiseptic dye. As a result, CV is the waste product of many of the processes which contaminate the water and also traverse through the food chain, leading to biomagnification. CV is carcinogenic and has been classified as a recalcitrant molecule since it is non-biodegradable, and can persist in a variety of environments. CV can easily interact with negatively charged cell membrane surfaces and can enter into cells and concentrate in the cytoplasm. CV can cause moderate eye irritation, painful sensitisation to light, heartbeat increase, vomiting, shock, cyanosis, jaundice, quadriplegia, and tissue necrosis. Therefore, it is essential to remove CV from industrial effluents before it is discarded into water bodies. CV has complex aromatic structure and is resistant toward chemicals, heat and light, and its biodegradation is a slow process. This review article provides the compilation of the information about CV, its classification and toxicity, various treatment methods, and its removal through different environmentally safe adsorbents. There are ample of biosorbents available for CV removal. One of the objectives of this review article is to organize the scattered available information on CV removal through adsorption using biosorbents including chitosan, surfactant-modified chitosan beads, NaOH-modified rice husk, grapefruit peel, coniferous pinus bark powder (CPBP), de-oiled soya, water hyacinth, peanut hull waste biomass, carboxylate-functionalized sugarcane bagasse, etc. The effectiveness of various adsorbents under different process parameters (such as solution pH, initial dye concentration, adsorbent dosage, and temperature) and their comparative adsorption capacity towards CV adsorption has also been presented. This review paper provides the understanding for further modification of the present adsorbents to improve their adsorption capacity and make them feasible for the field applications.

Keywords: adsorption, adsorbents, crystal violet, bio-polymers

1. INTRODUCTION

Currently, there are considerable environmental and ecological problems present in the world. The quality of drinking water and water for human consumption, irrigation has become a major concern for the public authorities, national and international bodies. Water pollution includes any change in water composition which is troublesome or harmful to human use. This modification of composition can be caused by release of toxic compounds or due to metal pollution. Specially, water contamination by dyes poses a serious problem, because of their high toxicity and non-biodegradability. The presence of dyes in the receiving environment is an undesirable situation in terms of aesthetics even if they are in very low concentrations. Furthermore, the dyed wastewater reduces the light transmission of the aquatic environment and affects adversely photosynthetic activities (Argun et al. 2017).

This type of pollution occurs due to the extensive use of dyes in several fields, including dyeing and printing on fiber and fabrics of all kinds, dyeing leather and furs, tincture of paper and parchment, dyeing of rubbers, sheets and plastics, dyes for all painting techniques, preparation of lime colors for precolorations and coatings on buildings, dyes for printing wallpapers, preparation of inks, coloring of foodstuffs, dyes for medicinal and cosmetic uses, etc. The increase of pollution due to industrial wastewater loaded with textile dyes, especially, crystal violet (CV), has created a major concern in developing countries. The importance of control of water pollution has steadily increased in recent years. A number of researchers are interested in developing new technologies and processes for the treatment of dyed wastewater. Various researches are focused on the creation of new types of adsorbents capable of eliminating these dyes. Many pollutants are resistant to the ordinary treatment processes and they persist in the environment for long time (Mani and Bharagava, 2016).

After implementing the increasingly strict regulations, some dyes have become the most worrying environmental pollutants which attract the attention of several research teams in environmental fields. Many researchers are working in the field of treatment of synthetic textile wastewater contaminated by different dyes. Several reported studies involve adsorption in batch (Miyah et al. 2017; Karim et al. 2018; Sun et al. 2016; Alizadeh and Mahjoub 2017; Patil et al. 2011; Smitha et al. 2012; Chowdhury et al. 2013; Lairini et al. 2017) or column mode (Chowdhury et al. 2013; Nidheesh and Gandhimathi 2018), adsorption on activated carbon (Jayganesh et al. 2017), degradation by Fenton (Su and Wang,

2011), gamma irradiation (Rehman et al. 2017), biodegradation by bacteria (Parshetti et al. 2011), separation by membrane filtration and ion exchange, coagulation-flocculation (Liu et al. 2018; Liu et al. 2018; Li et al. 2018), nanofiltration, photodegradation (Gupta et al. 2006) and electrocatalytic oxidation (Pillai et al. 2011).

Among all the technologies involved, some are associated with various problems such as excessive time requirements, high costs and high energy consumption, except for adsorption. The cost for adsorption process is manageable (Slokar and Le Marechal, 1998). So it can be said, that the adsorption is one of the viable economical techniques for the removal of CV from wastewater. The adsorption process in batch or column mode can be considered as an effective and widely used process due to its simplicity, moderate operational conditions and economical feasibility. Due to the high cost of activated carbon and other commercial adsorbents, viz., zeolites, composite, lignocelluloses and natural minerals, and due to the secondary waste generation by these adsorbents (Keyhanian et al. 2016), there is the need of the development of new sorbents effective in treatment and which are allies of the environment. Recently, numerous approaches have been adopted for the development of cheaper, more effective adsorbents which are suitable for laboratory conditions as well as for chemical industries. Among these, the eco-friendly adsorbents, such as clay minerals, coir pith, wood powder, lignin wood, saw dust, various agricultural waste materials like bagasse, peels of banana and orange, peanut hull etc. are the investigated ideal alternative for the expensive methods available for removing dyes from wastewater (Smitha et al. 2012). Also, polysaccharides, such as chitin, cellulose, chitosan, algae, alginate and its derivatives deserve particular attention for the treatment of dyes containing wastewater. All these sorbents can be used as finding cheap and effective alternatives for adsorption process. These adsorbents are considered as a renewable source that can replace petroleum derivatives and other sorbents. Due to their biological origin, these sorbents are classified as environmentally safe biosorbents.

Hence, in this review, we report the compilation of the information about CV such as structure, applications, toxicity, treatment methods and justified the choice of the effective method of treatment, and present its adsorption through different environmentally safe adsorbents. There are ample of biosorbents available for CV removal but we must mention the factors affecting the biosorption under different conditions (pH of the solution, initial concentration of dye, adsorbent dosage and temperature, phenomenon of coexistence, etc.). This paper is helpful for understanding the subsequent modifications of the adsorbents to improve their adsorption capacity and to propose other types of modification to make it useful for industrial applications.

2. DYES IN GENERAL WITH A SPECIAL REFERENCE TO CRYSTAL VIOLET

Dyes are classified into three categories depending on the charge they carry, for example, (a) anionic: acid and reactive dyes; (b) cationic: all basic dyes, and (c) nonionic: disperse dyes (El Qada et al. 2008). Colored dyes contain different organic compounds, which constitutes chromophores (NR_2, NHR, NH_2, COOH and OH) and auxochromes (N_2, NO and NO_2) (Gupta et al. 2009). Here, this paper is a detailed description about the CV, which is a cationic dye.

2.1. Chemical Formula, Structure and Properties of CV

Crystal violet (CV or BV10) having formula N, N, N^1, N^1, N^{11}, N^{11}-hexa-methyl-para-rosaniline is a triphenylmethane dye. CV is also known as gentian violet which is its impure form and the IUPAC name of CV is Tris (4-(dimethylamino) phenyl) methylium chloride. It has one dimethylamino group on each phenyl ring (Mani and Bharagava, 2016) and it has a triangular molecular structure. CV (hexamethyl pararosaniline chloride) is a basic dye with molecular formula $C_{25}H_{30}N_3Cl$ (Mona et al. 2011), which has a blue-violet color in appearance. Figure 1 shows the structure of CV.

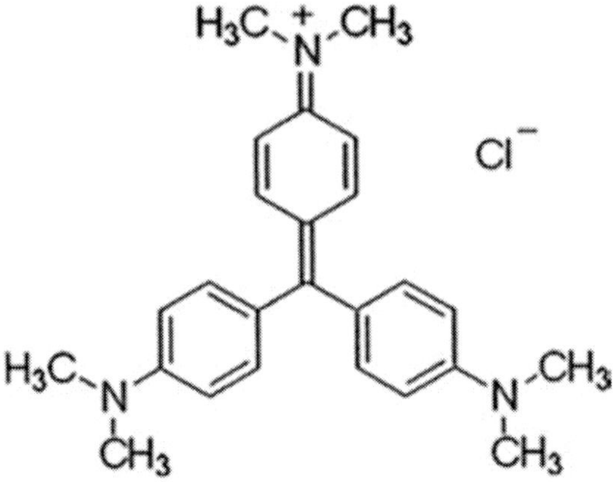

Figure 1. Chemical structure of crystal violet, molar mass: 408g mol^{-1} and λ_{max}: between 584-594 nm.

2.2. Source of CV Contamination in Environment, and Toxicity and Carcinogenic Effects of CV

Due to the application of CV in several industries (Table 1) it causes high toxicity to many industrial wastes. This results in generating considerable amount of colored wastewater which is discharged into the environment. Approximately 12% of synthetic dyes are considered to be lost while manufacturing and processing the products. Near about 20% of the resultant color is discharged into the environment due to the limitation of treatment processes in the industries (Essawy et al. 2008). About 4×10^4–5×10^4 tons of dyes are emanating to the water systems due to improper processing and dying methods from industries (El-Bindary et al. 2014). When CV is released in water bodies without adequate treatment, it blocks the sunlight penetration which causes the hindrance in the photosynthetic activity of aquatic plants. This ultimately causes the reduction in dissolved oxygen content and thus, finally disturbs the normal life process of the aquatic flora and fauna (Ajao et al. 2011; Cunningham et al. 2001).

Table 1. Fields of applications of CV

Fields	Applications
Biomedical engineering	• Biological activities (antibacterial, antifungal, and anthelmintic properties); (Adams and Rosenstbin, 1914) • Dentistry, and is known as "pyoctanin" (or "pyoctanine") • Marking the skin for surgery preparation and allergy testing • To stain tissue in the preparation of light microscopy sections in forensic (Docampo and Moreno, 1990). • Impetigo, used primarily before the advent of antibiotics, but still useful to persons who may be allergic to penicillin.
Cosmetics and toiletries	• Temporary hair colorant • Dermatological agent • Dyeing of cosmetics • Acrylic
Chemical industry andcommercial textile operations	• Biological stain • Dyeing cotton, nylon, polyacrylonitrile-modified nylon and wools • Dyeing textile, leather, paint, plastics, paper (navy blue and black) • Ball-point pens and ink-jet printers. • Coloring agent; such as fertilizers, detergents, and leather jackets. • Histological stain in Gram's method for classifying bacteria. • Used to avoid UV-induced DNA destruction when performing DNA cloning in vitro
Food industry	• Dye often used in the food industry
Pharmaceutics	• Dermatological agent

The dye CV also causes air pollution, ecosystem pollution and several negative effects on the environment when it is released in the air at a certain value. Human being if exposed to CV may have eye or skin irritation. CV also adversely affects the seed germination. It may be regarded as biohazard substance because of its toxicity and hazardous impacts on the human life. It may also cause the respiratory and kidney failure in extreme conditions (Ahmad, 2009; Mittal et al. 2010; Saeed et al. 2010). Fish tissues can readily absorb CV from dyed water where CV can be reduced metabolically to the leucocrystal violet (Mani and Bharagava, 2016). For all the above reasons we are interested in searching the treatment methods for this harmful dye using new cost-effective adsorbents.

3. ADSORPTION METHODS

Adsorption is among the physical treatment methods and it has recently become an important technique because of its greater efficiency in removing pollutants compared to other treatment methods. It is considered as an efficient and widely used process because of its simplicity, its moderate operational conditions, low cost and economic profitability. It is an effective method for the treatment of a wide range of heavy metals and dyes coming from wastewater. We can define the adsorption or sorption as a process by which contaminants present in one phase (usually liquid) accumulate on the surface of another phase (usually liquid or solid, beads particles, powders etc.). The degradation of dyes is difficult because of their complex aromatic molecular structures (Mani and Bharagava, 2016). Many techniques of depollution have been developed in recent years. These can be grouped into physical, chemical and biological treatment methods. It is worth mentioning the processes, their advantages and disadvantages for better understanding the applicability of the processes (Table 2). Adsorption has been found to be superior to other techniques because of flexibility and simplicity of design. Adsorption also does not result in the formation of harmful substances. The efficiency of the adsorption process is related to the properties of adsorbates and adsorbents, and it depends on the nature of their constituents. There are three types of adsorption based on the interactions involved. If the interaction between adsorbate and adsorbent is due to weak van der Waals forces it is called physisorption and process is reversible. On the other hand, if interaction is due to the chemical bonds, the adsorption process is called chemisorption. Chemisorption occurs only as a monolayer unlike physisorption. The third case is as a combination of chemisorption and physisorption. If there are favorable conditions, the two processes can take place simultaneously and it is called physico-chemisorption. The advantages, disadvantages and constraints of adsorption process used for dye removal are given in Table 3.

Table 2. Advantage and disadvantage of some selected methods of treatment for dye removal

Methods		Technology	Advantages	Disadvantages	Ref.
Conventional treatment processes	Chemical treatment methods	Coagulation and flocculation	Elimination of insoluble dyes. Simple and economically feasible	High sludge production, handling and disposal problems	(Kanamadi et al. 2006)
		Photochemical processes	No sludge production. Breakdown of compounds, and are non-hazardous	Formation of by-products. Economically less feasible because of high electricity costs	(Forgacs et al. 2004; Gogate et al. 2004; Robinson et al. 2001)
		Sodium hypochlorite (NaOCl) treatment	Initiates and accelerates azo bond cleavage	Release of aromatic amines	(Robinson et al. 2001)
		Ozonation	No sludge generation. No alteration of volume in the gaseous state that is why it is applied in gaseous state	Operational cost is very high, half-life is short (20 min)	(Dawood and Sen, 2014; Salleh et al. 2011)
		Electrochemical destruction	Breakdown of compounds, and are non-hazardous	Economically less feasible because of high electricity costs	(Robinson et al. 2001)
		Fenton reaction (oxidation system based on the Fenton's reagent)	Low-priced reagent and efficient procedure. Rapid process	Disposal issues and sludge production. High energy cost, chemicals required	(Salleh et al. 2011; Dawood et al. 2014)
	Biological treatment methods	Anaerobic degradation	By-products can be used as energy resources	Under anaerobic conditions require more treatment and yield of methane and hydrogen sulphide	(Salleh et al. 2011; Dawood et al. 2014)
		Aerobic degradation	Operational cost is low, and effective in removal of azo dyes	Provide suitable environment for growth of microorganisms and very slow process	

Table 2. (Continued)

Methods		Technology	Advantages	Disadvantages	Ref.
	Degradation mechanism by	Bacteria	Bacteria type-dependence CV resistant to biodegradation in the environment and the most important environmental factor affecting the CV biodegradation was the pH.	Choice of bacteria because many of them, which have been used to degrade CV, have been shown to be toxic to many other microorganisms.	(Michaels and Lewis, 1986)
		Fungi and their enzymes, yeasts, actinomycetes	Depends on the choice of the fungus Economically attractive, publicly acceptable treatment	Slow process, necessary to create an optimal favorable environment, maintenance and nutrition requirements	(Kanamadi et al. 2006)

Methods		Technology	Advantages	Disadvantages	Ref.
Established recovery	Physical Treatment methods / Membrane-filtration processes	Nano filtration	Separation of organic compounds of low molecular weight and divalent ions from monovalent salts	Membrane filtration removes all dye types. Produce a high-quality treated effluents but high pressures, expensive, incapable of treating large volume. Limited life-time before membrane fouling occurs The cost of periodic replacement must thus be included in any analysis of their economic viability Also, insufficient quality of the treated wastewater for ultra filtration and microfiltration processes Clogging of the pores which makes it an inefficient process	(Kanamadi et al. 2006; Ghayeni et al. 1998; Watters et al. 1991; Gupta et al. 2009)
		Ultra filtration and microfiltration	Low pressure		

Table 2. (Continued)

Methods		Technology	Advantages	Disadvantages	Ref.
Processes		Reverse osmosis	Removal of all mineral salts, hydrolyzes reactive dyes and chemical auxiliaries	High pressure	(Ghayeni et al. 1998)
		Electrodialysis and reverse electrodialysis	Operation is automatic, and maintenance is simple	Most organic contaminants are not eliminated
Water undergoing electro-dialysis must be pre-treated to reduce turbidity
the system must be cleaned regularly
process consumes a lot of electricity	
		Ion exchange method	Adsorbent can be regenerated and therefore used for several times and no loss of sorbent	Only effective for certain type of dyes
Economic constraints, not effective for disperse dyes	(Robinson et al. 2001; Kanamadi et al. 2006)
Processes		Activated carbon method	Suspended solids and organic substances. The most effective adsorbent, high capacity, produce a high-quality treated effluent	Cost of activated carbon, ineffective against disperse and vat dyes, regeneration is expensive and results in loss of the adsorbent, non-destructive process	(Mani and Bharagava, 2016; Kanamadi et al. 2006; Slokar and Le Marechal, 1998)

Table 3. Advantage and disadvantage of using biosorption/adsorption processes in batch mode and column mode for removal of dyes

Advantages	Disadvantages	Constraint
• The dye removal efficiency can be increased by adsorbent structure modification • Highly selective adsorbent can be produced and using bio-adsorbents cost of the process can be reduced • Easily manageable in batch mode and dynamic column • Regeneration is not necessary • High adsorption capacity for all dyes • Can be used for several adsorption/desorption cycles with a recovery rate of 100-80% of pollutants • Good physico-chemical resistance, and the desorption gives a material close to the initial state for an efficient reuse without any risk	• High cost of adsorbents and for bioadsorbents requires chemical modification • Unexpected behavior for some type of bioadsorbents that does not allow researchers to justify the high adsorption capacity • Low surface area for some adsorbents	• Mechanisms complex and not fully understood • The choice of special and effective adsorbents for each type of pollutant • Need modifications to be easily used in industrial wastewater treatment and also for other complex condition outside the laboratory • Nature, complicated structure and the dimensions of pollutants this influences the adsorption/desorption mechanism

3.1. Factors Affecting the Biosorption

The effectiveness of removal process depends on the nature of the adsorbents like structure of adsorbent, exiting active site, nature of contaminants, the mechanisms involved in the adsorption etc. Many factors are of crucial importance to improve the efficiency of adsorption process. We can classify them as follows:

- The origin of the sorbents or polymer
- The degree of polymerization (DP)
- Special proprieties such as degree of acetylation (DD) in case of chitosan and M/G ratio in the case of alginate
- The amount of basic material in the preparation reaction and the percentage of grafted functions or the type of the cross linking agent.
- The shape and size of the particles in the adsorbents and the size of the pores.
- The structure of the adsorbent and the types of existing sites in their surface.
- The physical form of the adsorbents or the nature of the beads (dry to wet) and also drying method (controlled drying such as lyophilization and supercritical CO_2 drying).
- The nature of the dye and these properties are affected by the size of the compound to be trapped (surface, volume).
- pH plays a very important role in the mobility of dyes.
- The dosage of an adsorbent increases the number of active sites that vary according to the objectives of the treatment (decontamination or dye recovery).
- Temperature and the study of the effect of this parameter gives useful information on the thermal nature of the adsorption and its spontaneity
- Time and stirring speed, in some cases if the stirring speed increases, it increases the rate of elimination by minimizing their

resistance to mass transfer. But, this operation can damage the physical structure of the adsorbent.
- The initial concentration of metal ions.
- The phenomenon of competition between dyes (if the solution is complex, containing pollutants in coexistence with another pollutant, competition can reduce the elimination of the target pollutant).

3.2. Sorption/Biosorption Mechanisms

The mathematical modeling of kinetics and isotherms provide important information to understand the mechanism of adsorption. This is crucial and helps in selecting the desorption strategy. Despite of their importance and the large number of documents devoted to the elimination of dyes by different materials, maximum interest has been paid on the evaluation of adsorption performance rather than to understand the mechanism of sorption. The kinetics of sorption varies according to the materials used in the adsorption process. After a certain time, the adsorption becomes slow and finally it stabilizes with the increase in contact time. Generally, there are three main stages in the mechanism of sorption of dyes on the adsorbent viz., (a) transport of the pollutant from liquid to solid phase, (b) adsorption on the surface of the material, (c) transport in the particles of the adsorbent. Therefore, the physico-chemical mechanisms responsible for the retention of pollutants in the adsorbent matrix by adsorption can be: physisorption, chemisorption, complexation, precipitation and substitution.

In the cases of polysaccharide materials, due to the complex structure of these adsorbents and their specific characteristics such as presence of chemical groups, small specific surface area and low porosity, the sorption mechanisms are even more complex and anonymous. Crini (2005) mentioned that there are several types of interactions, which can act simultaneously. Some of these interactions that are reported can be physical adsorption, electrostatic interaction, the covalent and hydrogen bonds, the ion exchange, complexation, Lewis acid-base type of interactions, hydrophobic interactions and precipitation, etc. It is believed that the combination of these interactions can occur during the adsorption of polysaccharide-based materials.

The choice of linear or non-linear regression method and the model used in the kinetic or isothermal modeling of the data provides the understanding of the sorption mechanism. There are theoretical models with two parameters and with three parameters. The comparison of the results obtained is done by coefficient of correlation R^2 and/or EV (estimated variance) or other method of estimation of error obtained between the experimental value and the value obtained by modeling. Zhou et al. (2014), mentioned that the valence forces due to the exchange of electrons between the adsorbent and adsorbate plays a key role during adsorption which depends on the number of adsorption sites available on the surface of the adsorbent (Zhou et al. 2014). The forces for adsorption include van der Waals forces, electrostatic interactions, π–π interactions, and hydrogen-bonding forces, with electrostatic interactions as the main force (Khan et al. 2013; Mori et al. 2013). Some of the important works focused on the adsorption mechanism is summarized in Table 4. These include the work of Eskhan et al. (2018), the grafting of alginate with poly (styrene-co-maleic anhydride) (PSMA) establishes electrostatic attraction and π-π stacking interactions with the aromatic rings of the cationic dye improving the removal percentage of the dye (Eskhan et al. 2018a). Also, Kumar et al. (2011), confirmed that the structure of the dye is a prominent factor for biosorption. The complexation between CV and treated ginger waste (TGW) can take place through the weak and strong forces (Kumar and Ahmad, 2011). The weak interactions occur due to the van der Waals forces while the strong interactions occur due to:

- hydrogen bonding interaction between the nitrogen containing amine groups of CV and TGW surface

- hydrophobic–hydrophobic interactions between the hydrophobic parts of CV and TGW.
- electrostatic interaction between the cationic dye (due to the presence of $^+N(CH_3)_2$ group) and negatively charged TGW surface in basic medium (Kumar and Ahmad, 2011).

Saeed et al. (2010) studied the application potential of grapefruit peel (GFP) as dye sorbent and confirmed that at a pH >4, the carboxylic groups are deprotonated and bind the positively charged CV molecules. This confirms that the sorption of CV by GFP is ion exchange mechanism between the negatively charged groups present in the cell wall of GFP and the cationic dye molecule. Maximum sorption capacity obtained for the above process was 254.16 mg g^{-1}. Proposed mechanisms of CV adsorption is shown in Figure 2 (Saeed et al. 2010).

Although the presence of the amino (-NH$_2$) and the hydroxyl (-OH) groups on chitosan chains serves as the reaction sites but sometimes it needs surface modification to increase its adsorption efficiency. Sakkayawong et al. (2005) used chitosan (MMWC) beads cross-linked with ethylene glycol diglycidyl ether for the sorption of cationic dyes. Hydroxyl group (-OH) in chitosan could adsorb the basic dye via covalent and hydrogen bonding, following similar mechanism as that of the adsorption of cellulose polymers with reactive dyes (Sakkayawong et al. 2005). For alginate, cellulose and other similar materials generally adsorption occurs by interaction between cationic dyes on the basic surface. But due to the method of surface modifications like grafting, physical or chemical cross-linking and others, it is possible to improve the characteristics of the materials.

Table 4. Some of the proposed mechanism for CV (or dyes of the same family) removal onto natural materials such as chitosan, clay, alginate and others

Material	Proposed mechanism	Illustration of mechanism	Reference
Chitosan–graphite oxide modified polyurethane	Weak interactions due to van der Waal forces, strong interactions due to hydrogen-bonding, interactions between nitrogen-containing amine groups of CV and the surface of sorbents		(Qin et al. 2015)
Fe@GAC (flyash/bentonite (2:1)) Fe@GAR (flyash/bentonite /coke/iron)	CV removal by Fe@GAC and Fe@GAR included adsorption and reduction		(Liu et al. 2018)
Tomato plant root (TPR)	Interaction between cationic CV on the basic surface of the TPR powder		(Kannan et al. 2009)

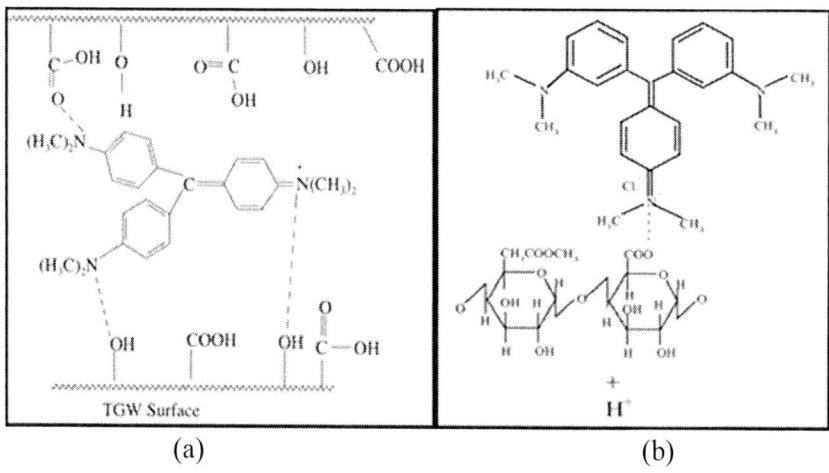

Figure 2. (a) Proposed mechanism for the biosorption of CV onto treated ginger waste (TGW) (Kumar and Ahmed 2011), (b) Proposed ion exchange mechanism between a proton of grapefruit peel and CV (Saeed et al. 2010).

4. RECENT REPORTS ON VARIOUS ADSORBENTS USED IN CV REMOVAL

Various adsorbents have been tested for the decontamination of CV in aqueous solutions. Sun et al. (2016) reported that, *Bacillus amyloliquefaciens* biofilm is higly efficent for removal of CV. Moreover, the determined maximum capacity for the CV adsorption was 582.41 mg/g (Sun et al. 2016). Kulkarni et al. (2017) studied the equilibrium, kinetics and thermodynamics of adsorption using water hyacinth and reported rapid uptakes of CV in 40 min, followed by a slow uptake over the next 5 h in the range of temperature at 300–323 K. Maximum capacity for the CV adsorption obtained was 322.58 mg/g (Kulkarni et al. 2017). Kannan and Sundaram (2001) reported the adsorption capacity of activated carbon prepared from straw to be 472.10 mg/g (Kannan and Sundaram 2001). In the work of Ruan et al. (2018) it was shown that, rGO/Fe/Ni composites (reduced-graphene-oxide bimetallic Fe/Ni nanoparticles) were able to get the maximum adsorption capacity of 2000.00 mg/g (Ruan et al. 2018).

Table 5. Recent reports on various adsorbents used in water treatment in batch and column mode

Material	pH	Temp. (°C)	Dose (g/L)	Equilibrium time (min)	q_m (mg/g)	Mechanisms of adsorption	Reference
Eco-friendly activated carbon from *sargassm wightii* sea weeds	7.0	30 ± 0.5 °C	-	60	21.05	-	(Jayganesh et al. 2017)
Potato peels (*Solanum tuberosum*)	6.0	30°C	0.5	20	555	-	(Lairini et al. 2017)
Tomato plant root	3.85	30-40°C	-	60	94.34	Chemisorption	(Kannan et al. 2009)
Natural materials (mangrove plant, mango, tamarind, teaktree, almond tree)	7.0	30 °C	1.0	60	142.80-250	-	(Patil et al. 2011)
$NiFe_2O_4$ magnetic nanoparticles treated with sodium dodecyl sulfate($NiFe_2O_4$–SDS)	5.5	25°C	-	25-35	21.46	Physical adsorption	(Alizadehet al. 2017)
Coating of biochar with magnetic Fe_3O_4 nanoparticles	6.0	40°C	1.0	240	349.40	Chemisorption	(Sun et al. 2015)
Powdered mycelia biomass of *Ceriporia lacerata* P2	natural pH	20°C	0.20	300	239.25	Hydrogen bond and ion exchange	(Lin et al. 2011)
Banana leaves, stem and stalk	7.0	25°C	-	180	93.2-95.4	-	(Kartina et al. 2018)
Coal bottom ash (CBA)	5.0	25 ± 2°C	-	12	17.7	Chemical nature	(Bertolini et al. 2013)
Zeolite from coal bottom ash (ZBA)			-		19.6		
NaOH-modified rice husk	$C_0 = 100$ mg L^{-1}, flow rate = 22.88 mL min^{-1} and bed height 18.75 cm in 12 h				70.275 (Thomas model)	-	(Chowdhury et al. 2013b)

Material	pH	Temp. (°C)	Dose (g/L)	Equilibrium time (min)	q_m (mg/g)	Mechanisms of adsorption	Reference
CMRS (Citric acid modified rice straw)	8.0	20°C	30.0	360	90.82	Chemical-ion exchange mechanism	(Chowdhury et al. 2013a)
Cucumis sativa activated carbon	7.0	27 ± 2°C	4.0	90	34.24	-	(Smitha et al. 2012)
Surfactant modified magnetic nano adsorbent	6.0	-	0.25	10	166.67	Physiosorption process due to electrostatic or van der Waals attraction	(Muthukumaran et al. 2016)
Coniferous pinus bark powder (CPBP)	8.0	30°C	-	120	32.78	-	(Ahmad, 2009)
Starch composite with peanut hull	>7	50°C		60	101.50	-	(Tahir et al. 2017)
Polypyrrole composite with Peanut hull			0.5		89.10		
Polyaniline composite with Peanut hull					90.20		
Chemically modified phoenix Tree leaves	6	20°C	0.6-1.8 g/L	660	510.3	ion exchange	(Ren et al. 2015)
Modified bambusa tulda	7.0	25°C	10.0	60	20.84	chemical adsorption mechanism	(Laskar and Kumar 2018)

Table 5 is the summary of the parameters used to get the maximum adsorption by various researchers.

5. USE OF LOW-COST ADSORBENTS IN THE SORPTION OF CV

In recent years, the development of commercial applications for chitin, chitosan, algae and alginate has progressed. The emphasis on environmental friendly technology has spurred interest in these biopolymers, which are versatile, biodegradable and less toxic than the synthetic materials. The reactive groups (hydroxyl, carboxyl, acetamide or amine functions) in the chains of these biopolymers have justified the interest of their use in the treatment of industrial wastewater. Many approaches have been made to modify these adsorbents in order to adapt them to the requirements of the chemical industry.

These two polysaccharides, chitosan and alginate, have many unique advantages and characteristics, such as their abundance, non-toxicity, biocompatibility, reactivity, biodegradability and their effectiveness for the treatment of dyes as well as metal contaminants.

5.1. Chitosan and Chitosan Derivatives for the Removal of CV

Chitosan is produced from chitin, which is rich in reactive functional groups (Figure 3). Chitosan can remove various types of pollutants such as metal ions and dyes (especially anionic dyes). Chitosan is considered as the second most abundant biopolymer after cellulose in the world (Guibal et al. 2002; Hamutoğlu et al. 2012). The properties manifested by chitosan are versetile. It is suitable for chemical modification. It can be made to various shapes, and above all it is a cationic polymer in acidic medium. It has a polycationic structure where-NH_3^+ reacts with anionic surfaces. Semi-stiffness and high viscosity of chitosan is due to β (1-4) side binding.

Apparent pk$_a$ is 6.5 and it aggregates in solutions having a pH above 6.0, it is therefore only soluble in acidic solutions in the pH range 1-6.

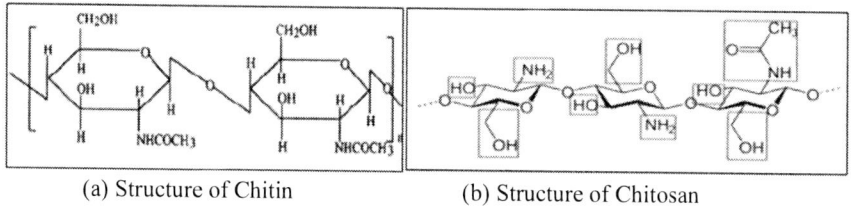

(a) Structure of Chitin   (b) Structure of Chitosan

Figure 3. Chemical structures of chitin and chitosan, and illustration of the possible reaction sites on chitosan surface.

The -NH$_2$ group of chitosan is very important in the adsorption because this function leads to the formation of complexes with various metal ions and various types of dyes such as the anionic dyes. This function provides selective reactions in C-2 position of D-glucosamine so as to modulate the properties of the chitosan and to obtain new physical properties. Also, the –OH group worn by the C-3 is certainly the stabilizing group fixing metal ions or dye molecules without being the main trigger. For all these reasons, chitosan has become an adsorbent that can be used for removal of pollutants. The use of chitosan and its derivatives in the field of dye removal is not less important than that of heavy metal removal. It has shown the promising results for removal of dyes from aqueous solutions (Shouman et al. 2012).

Several types of dye removal are studied by chitosan. For example, Ling et al. (2011) studied the biosorption of dye by chitosan. The adsorption followed Langmuir model with the maximum adsorption capacity of 31.65 mg/g, and the process followed ion-exchange mechanism i.e., chemisorption. Authors reported that, adsorption increases with the increase in degrees of deacetylation of chitin (Ling et al. 2011). Concerning the study by Qin et al. (2015), the adsorption of CV on chitosan-graphite oxide modified polyurethane, with a fixed dose of 50 mg/mL, temperature of 35°C and a pH of 8, the maximum capacity recorded was 64.93 mg/g(Qin et al. 2015). Kumar et al. (2017) studied the effectiveness of composite TLAC/Chitosan for removal of CV in aqueous solutions at pH=9, dose 0.4 g/L, 303 K at 30 mg/L of CV. The follow-up of the literature report led us to select the important adsorption study of CV, and to summarize several parameters which are related to the adsorption of the CV using chitosan. Guibal et al.(2003) and other researchers confirmed that, increase in the degree of deacetylation generally gives an increase in sorption capacity for anionic dyes due to the availability of protonated amine groups (Guibal et al. 2003; Saha et al. 2005). The presence of protonated amino groups in chitosan contributes towards better sorption capabilities of chitosan as compared to chitin by allowing ionic bonding between the amino groups and the dyes (Wong et al. 2008). Other related works are summarized in Table 6, which confirms the effectiveness of using chitosan as adsorbent for CV recovery/decontamination. However, any comparison remains relative since the operating conditions are often different.

5.2. Alginate and Alginate Derivatives as Environmental Friendly Adsorbents for the Removal of CV

Alginate is a linear polysaccharide and binary copolymer of D-mannuronic (M) and L-guluronic (G) acids linked together by 1-4 glycosidic bond. Alginates are derived from seaweeds/brownalgae. The M/G ratios depend on the source or species, the growth conditions, season and extraction method. The variation of M and G along the alginate chain determines its physical properties and reactivity towards pollutants. Also, alginates with a high content of G are stronger than others. Figure 4 shows the structure of alginate, where –COOH and –OH groups are responsible for the fixation of pollutants and most physicochemical changes to its structure.

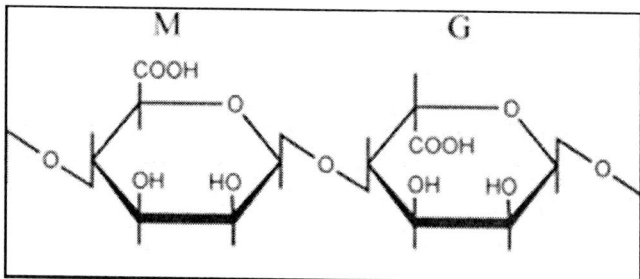

Figure 4. Structure of alginate.

It is one of the most varied polymers for its industrial exploitation and has found many applications in the fields of depollution, agri-food or textile engineering. Ca^{2+} is the most commonly used cation for inducing alginate gel formation. There are two methods of reticulation of alginate by Ca^{2+} which are "diffusion" method and the "internal control" method. Diffusion method produces gels having a difference in concentration of Ca^{2+} ions in the thickness, whereas the internal setting control gives gels with uniform concentrations of ions throughout the product (Skjak-Braek et al. 1989). The grafting and crosslinking are common ways to improve sorption properties by adding various functional groups for enhanced target dye selectivity and also improved chelation or complexation properties. For alginate and chitosan, the chemical modification is used as a tool to achieve the objectives viz., (1) improve the existing properties, and (2) introduce completely new properties that do not exist in unmodified/parent alginate and chitosan. Several materials like glutaraldehyde (GLA), epichlorohydrin (EPI), ethylene glycol diglycidyl ether (EGDE), as well as poly (vinyl alcohol) (PVA), poly (acrylic acid) (PAA), and polyethylenimine (PEI) have been used as crosslinking agents for the preparation of chitosan and alginate-based materials which are used in sorption of dyes. Other water-soluble crosslinking agents have been proposed by several researchers, such as sodium trimetaphosphate, sodium tripolyphosphate, phosphorus oxychloride or carboxylic acids, etc. Some of the modified forms are compiled in Table 6.

Table 6. Some recent studies for CV removal by low cost adsorbents based on chitosan and alginate

Material	Time (min)	pH	Temp. (°C)	Dose (g/L)	Rotation speed (rpm)	q_m (mg/g)	Model fitted	Reference
Acid activated bentonite clay (AB)	6 h	8.0	25	2.2	200	498.2	Langmuir	(Oladipo and Gazi, 2014)
alginate/bentonite beads (AAB)						229		
Alginate/acidactivated bentonite beads (A-AAB)						582.4		
Fe@GAC (flyash/bentonite (2:1)) Fe@GAR (flyash/bentonite/coke/iron)	24 h	9-12	20	-	-	95.24 123.45	Langmuir	(Liu et al. 2018)
Jujube shell NaOH-modified jujube shell	60	5.46	50	0.2 0.1	-	59.84 288.18	Langmuir	(El Messaoudi et al. 2017)
Ceriporia lacerates P2	5 h	Natural pH	20	0.20		239.25	Langmuir	(Lin et al. 2011)
Treated ginger waste (TGM)	150	6.2	50	0.05	80	277.7	Langmuir	(Kumar and Ahmad, 2011)
Magnetite alginate	30	7.0	25	0.05	-	37.5	Langmuir	(Elwakeel, El-Bindary et al. 2017)
Clay type; Montmorillonite K10 0.25 M acid-treated montmorillonite (Mt1) 0.50 M acid-treated montmorillonite (Mt2)	180	5.9	30	0.4	-	370.37 384.62 400.0	Langmuir	(Sarma et al. 2016)

Material	Time (min)	pH	Temp. (°C)	Dose (g/L)	Rotation speed (rpm)	q_m (mg/g)	Model fitted	Reference
Modified cellulose	150	9	10- 50	0.25	190	118.2-304.9 169.9-218.8	Langmuir	(Zhou et al. 2014)
Magnesium-oxide coated bentonite (MCB)	200	6.5	35	2.0		496	Langmuir	(Eren, 2009)
Sawdust	-	-	25	1.0	-	341.0	Langmuir	(Chakraborty et al. 2005)
Alginate-fixed water hyacinth	200	10.0	30	10.0	150	300 mg/L q= 126.2	-	(Mahamadi and Mawere, 2013)
Calcium alginate hydrogel wasgraftedwith poly (styrene-co-maleic anhydride) synthetic polymer (PSMA)	-	5.0	40	-	-	109.9	Langmuir	(Eskhan et al. 2018b)
Clay of morocco	-	$pH_{PZC} = 9.2$	20	1.0	-	17.05	Langmuir	(Miyah et al. 2017)
Low-cost adsorbents made from waste of *Rapanea ferruginea* treated with ethanol (WRf) H_2SO_4-treated analog WRf/H_2SO_4	120		25 25 55 55	0.2	-	148 142 9 34	Langmuir	(Chahm et al. 2018)
Natural claymineral	180	7.0-7.5	20-50	0.5		330.0	Langmuir	(Omer et al. 2018)
Calcium alginate hydrogel grafted with poly (styrene-co-maleic anhydride)	100	-	40	4.6 g/L		109.9	Langmuir	(Eskhan et al. 2018a)

Table 6. (Continued)

Material	Time (min)	pH	Temp. (°C)	Dose (g/L)	Rotation speed (rpm)	q_m (mg/g)	Model fitted	Reference
Kappa-carrageenan (Car) and sodium alginate (Alg)	-	6.4	-	10.0	120	66.7	Langmuir	(Mahdavinia et al. 2013)
Nanocomposite hydrogels by incorporation of sodium montmorillonite (Na-MMt) nanoclay CarAlg/MMt 0.1						85.2		
Raw bentonite	90	7.0	30	0.5		5.67	Langmuir	(Laysandra et al. 2018)
Acid activated bentonite (AAB)						79.27		
Rarasaponin–bentonite						79.79		
Biocharfrom durian shells						32.40		
Activatedbiocharfrom durian shellswith KOH concentration (1.65 M)						51.48		
Superabsorbent hydrogel (SAHs) composed of (acrylicacid, sodium alginate and sodium humate) (AAc/NaAlg/SH)	1440	-	-	-	-	289.0	Langmuir	(Agnihotri and Singhal, 2018)
Chitosan	120	8.0	23-25	-	150	28.5	Langmuir	(Shouman et al. 2012)

Material	Time (min)	pH	Temp. (°C)	Dose (g/L)	Rotation speed (rpm)	q_m (mg/g)	Model fitted	Reference
Magneticchitosan adsorbent modified with EDTA (EDCMS)	720	7.0	25	-	-	227.3	Langmuir	(Ren et al. 2014)
O-carboxymethylchitosan cross-linked (OCMCTS)	240	8.0	30	-	120	227.27	Langmuir	(Sarkar et al. 2012)
Chitosan pyrrole composite with peanuthull Chitosan aniline composite with peanuthull	60	>7	50	-	-	93.50 105.20	-	(Tahir et al. 2017)
Chitosan (CSC) beads MMWC beads*	30	2-4	-	-	-	79.6 21.5	Langmuir	(Jassal and Raut, 2015)
Chitosan hydrogel beads	180	7.0	30	0.5	150	76.9	Langmuir	(Pal et al. 2013)
Poly(AA-co-VPA) hydrogel cross-linked with N-maleylchitosan	1440	7.0	25 55	0.05	-	60.53 64.56	Langmuir	(Nakhjiri et al. 2018)
Kaolin/chitosan/titanium dioxide ultrasonic-CTS-KL-TiO$_2$ (116.5 m^2/g) Conventional-CTS-KL-TiO$_2$ (4.95 m^2/g)	180	-	25	1.0	250	111.11 88.496	Langmuir	(Vardikar et al. 2018)

*MMWC medium molecular weight chitosan beads cross-linked with ethylene glycol diglycidyl ether.

Table 7. Advantages and disadvantages of some of the eco-friendly adsorbents used for the removal of dyes from water

Adsorbents	Advantages	Disadvantages
Chitin	• High specific area, easy functionalization, high crystallinity, high modulus, low-cost (15 –20 US $/kg).	• Need modification before use
Polysaccharides such as; Chitosan-based material and alginate-based material	• Low-cost natural polymer and environmental friendly (chitosan (16.5 –10 US $/kg), cross-linked-chitosan (5–10 US$/kg)) (Çifçi and Meriç, 2016; Grassi et al. 2012) • Easy extraction process from algae and easy deacetylation of chitin • A high bioavailability, very abundant material and widely available in many countries • Renewable resource, extremely profitable, environmental friendly and acceptable material from a public point of view • Versatile sorbent and rich surface with specific active sites for several types of pollutants (for alginate carboxyl groups and hydroxyl groups and for chitosan amino groups) • Ability to associate, through physical and chemical interactions, with a wide variety of molecules and shows fast removal kinetics • Extremely cost-effective and the adsorption process and the optimization of the sorption parameters decreased their costs • Versatile properties that can be monitored according to changing the parameters	• Variability of the characteristics of the polymer • Low affinity for basic dyes for chitosan and low affinity for acidic dyes for alginate • The sorption capacity depends on the origin of the polysaccharide and the degree of N-acetylation for the chitosan and the ratio M/G for the alginate • Requires chemical modification to add other special function to improve its performance for adsorption in some cases of dyes • Poor porosity and low specific surface area, for example chitosan 0.970 m^2/g (Vardikar et al. 2018) requires modification to improve the porosity. • pH, v/m and temperature dependence for dye solution and requires an optimization of the parameters

Adsorbents	Advantages	Disadvantages
	- Outstanding metal and dye-binding capacities compared to other low cost biosorbents - Easy regeneration if required and usable for several adsorption-desorption cycles	- Difficult to use some physical forms in column and pilot mode in the industry as the case of the fine powder in the columns and the need growing to find other effective forms for ease of use in the industry - Limitation of use in sorption columns (hydrodynamic limits and fouling of columns)
Clay-based materials	- Clay sorbent (bentonite: 0.05 –0.2; Red mud: 0.025; Clinoptilolite: 0.14 –0.29 US$/kg) (Çifçi and Meriç, 2016; Grassi et al. 2012) - Strong attraction to cationic and anionic dyes but greater adsorption capability for basic dyes comparable to acid dyes - Great efficiency raw clay minerals to uptake anionic dyes from a liquid solution. - A high bioavailability - Clay adsorption capacities depend on the net negative charge found on the clay minerals. - High cation exchange capacity and large pore sizes and high surface area - The size of clay particles plays an indispensable role in all inter-phase interactions in clays including ion-exchange and adsorption processes. - The molecular structure has effect on the adsorption of basic dyes, that can also play a vital role in the adsorption processes - Adsorption of CV by different types of clays is exothermic process	- The modified clays offer more efficiency for the adsorption of the dyes compared to the raw clays - Difficulty in recovering of clay particles from solutions after the adsorption process makes them even less attractive as adsorbents for industrial water treatment. - Regeneration of these sorbents is difficult. So chemical modification is required. - Low surface area of some types of clays and low efficiency of micro-pollutants treatment compared to activated carbon - Adsorption capacities can differ even though the dyes have the same charge. - Adsorption affected by extent of hydration, and their negatively charged layers

Table 7. (Continued)

Adsorbents	Advantages	Disadvantages
Cellulose-based materials	<ul><li>Easy functionalization in some cases,</li><li>Hydroxyl groups, and functional groups form derivatives</li><li>Large potential of future applications in some interesting fields like ultrafiltration, microfiltration, adsorption, trace metal detection</li><li>Very low price (~5-175 US $/kg) depending on the place of purchase and the type of cellulose (powder, fibers etc.), but generally cellulose is locally sourced directly from the nature or from the manufacturer (Hokkanen et al. 2016).</li></ul>	<ul><li>Modification method is expensive with a lot of disadvantages such as high consumption of chemicals, and necessity of using some harmful chemicals, as well as a complex synthesis pathway.</li><li>In case of nanocellulose the price is very volatile and may range from 50 US $ to 1080 US $/kg (Hokkanen et al. 2016).</li><li>Low adsorption capacities for some modified cellulose</li></ul>
Other eco-friendly adsorbents such as: Starch, peat, sawdust, fly ash, coniferous pinus bark, de-oiled soya, Aloevera, hyaluronic and derivatives, peanut hull waste biomass, sugarcane bagasse, etc.	<ul><li>Low-cost sorbents</li><li>Very abundant natural biopolymer</li><li>Environmental friendly</li><li>Available in many countries with these charchetrictics and compositions changes according to the country.</li><li>Different functional groups and good removal and recovery of wide range of dyes</li><li>Easy functionalization</li></ul>	<ul><li>Low adsorption capacity in its raw form</li><li>Usually needs chemical treatment before use</li><li>Low removal properties and selectivity compared to activated carbons</li><li>Required chemical modification</li><li>Required innovative methods to develop more efficient eco-friendly adsorbents</li></ul>

6. ADVANTAGES AND POSSIBLE DRAWBACKS OF USING ENVIRONMENTALLY SAFE BIOSORBENTS

The adsorption on the environmentally safe biosorbents and their derivatives is an inexpensive procedure in the decontamination of water, extraction and separation of the compounds. It is also an effective tool for protecting the environment. These materials become even more effective when combined with other selected compounds which improve their structural properties for desired applications. Generally, an effective adsorbent should meet several requirements, for example, efficiency for the removal of a wide variety of pollutants at an acceptable adsorption rate; high selectivity at different concentrations; granulometry offering a large area; strong physical resistance; regenerability if necessary; tolerance for a wide range of operating parameters; should be cheap. Table7 summarizes some of the advantages and disadvantages of using alginate, chitosan, cellulose, clay-based materials, and some other adsorbents in the removal of dyes. The table also shows the comparison of the effectiveness and the costs of materials containing chitosan, alginate with other low-cost adsorbents in the sorption of CV.

CONCLUSION

Many industries such as textile, paint, cosmetic, food industry, plastic and pharmaceutical industries use large quantities of water. It is important for the industries to treat and recycle/reuse their wastewater before throwing it into the environment. The available information about crystal violet (CV) reveals that, this dye has now become one of the most controversial compounds due to its negative effects and health hazards posed on the environment and on all living organisms. Thus, it is concluded that CV is a head strong pollutant, which possesses toxic effects on aquatic as well as on terrestrial ecosystems. It also acts as a carcinogenic agent. This review deals with the adsorption/sorption/biosorption of CV by environmentally safe adsorbents. Their advantages and disadvantages in the sorption of CV are identified. The methods of the synthesis of biosorbents and their contributions in the sorption of CV are noted and discussed.

Till date, a number of modifications have been tried to develop better quality adsorbents, which show higher adsorption capacity. Some of the raw and modified adsorbents with their optimized parameters and adsorption capacities are summarized here for the better understanding of the best adsorbents for CV removal from aqueous solution. On the basis of the review paper we can conclude that there are a number of adsorbents available which can be utilized for CV removal from industrial wastewater in cost effective way. Biosorption process for treatment of CV is environment friendly.

ACKNOWLEDGMENTS

Authors would like to thank Indian Institute of Technology Kharagpur (IIT Kgp), for providing financial assistance to Ms. Preeti Pal to carry out this research. We are also thankful to Université de Sciences et de la Technologie, Algérie, Laboratoire de Génie Chimique et de catalyse hétérogène and Ecole des Mines d'Alès, Centre des Matériaux des Mines d'Alès (C2MA), Cedex, France for providing financial assistance to Ms. Asmaa Benettayeb.

REFERENCES

Ajao, A. T., Adebayo G. B., and Yakubu S. E.,(2011). Bioremediation of textile industrial effluent using mixed culture of *Pseudomonas aeruginosa* and *Bacillus subtilis* immobilized on agar-agar in a Bioreactor. *J. Microbiol. Biotech. Res.*1(3), 50–56.

Adams, E.Q., Rosenstbin, L., (1914). The color and ionization of crystal-violet. *J. Am. Chem. Soc.,* 36(7), 1452-1473.

Agnihotri, S., Singhal, R., (2018). Synthesis and Characterization of novel poly (acrylic acid/sodium alginate/sodium humate) superabsorbent hydrogels. Part II: the effect of sh variation on Cu^{2+}, Pb^{2+}, Fe^{2+} metal ions, MB, CV dye adsorption study. *J. Polym. Environ.* 26(1), 383–395.

Ahmad, R., (2009). Studies on adsorption of crystal violet dye from aqueous solution onto coniferous pinus bark powder (CPBP). *J. Hazard. Mater.* 171(1-3), 767-773.

Alizadeh, N., Mahjoub, M., (2017). Removal of crystal violet dye from aqueous solution using surfactant modified $NiFe_2O_4$ as nanoadsorbent: Isotherms, thermodynamics and kinetic studies. *J. Nanoanalysis,* 4(1), 8–19.

Argun, Y., Karacali, A., Calisir, U., Kilinc, N., Irak, H., (2017). Biosorption method and biosorbents for dye removal from industrial wastewater: Areview. *Int. J. Adv. Res.* 5(8), 707–714.

Bertolini, T., Bertolini, T.C.R., Izidoro, J.C., Magdalena, C.P., Fungaro, D.A., (2013). Adsorption of crystal violet dye from aqueous solution onto zeolites from coal fly and bottom ashes. *Orbital-Electron. J. Chem.* 5(3), 179–191.

Chahm, T., Martins, B.A., Rodrigues, C.A., (2018). Adsorption of methylene blue and crystal violet on low-cost adsorbent: waste fruits of *Rapanea ferruginea* (ethanol-treated and H_2SO_4-treated). *Environ. Earth Sci.* 77(13), 508.

Chakraborty, S., De, S., Das Gupta, S., Basu, J.K., (2005). Adsorption study for the removal of a basic dye: Experimental and modeling. *Chemosphere.* 58(8),1079-1086.

Chowdhury, S., Chakraborty, S. and Das, P. (2013a). Adsorption of crystal violet from aqueous solution by citric acid modified rice straw: Equilibrium, kinetics, and thermodynamics. *Sep. Sci. Technol.* 48(9), 1339–1348.

Chowdhury, S., Chakraborty, S. and Das, P. (2013b). Response surface optimization of a dynamic dye adsorption process: A case study of crystal violet adsorption onto NaOH-modified rice husk. *Environ. Sci. Pollut. Res.* 20(3), 1698–1705.

Cifici, D.İ., Meriç, S., (2016). A review on pumice for water and wastewater treatment. *Desalin. Water Treat.* 57(39), 18131–18143.

Cunningham, W.P., Saigo, B.W. and Cunningham, M.A., (2001). *Environmental science: A global concern* (Vol. 412). Boston, MA: McGraw-Hill, New York, NY.

Dawood, S., Sen, T.K., (2014). Review on dye removal from its aqueous solution into alternative cost ef- fective and non-conventional adsorbents. *J Chem Proc Engg.* 1, 1-7.

Docampo, R., Moreno, S.N.J., (1990). The metabolism and mode of action of gentian violet. *Drug Metab. Rev.* 22(2-3), 161-178.

El-Bindary, A.A., Diab, M.A., Hussien, M.A., El-Sonbati, A.Z., Eessa, A.M., (2014). Adsorption of Acid Red 57 from aqueous solutions onto polyacrylonitrile/ activated carbon composite. Spectrochim. *Acta-Part A Mol. Biomol. Spectrosc.* 124 (2014), 70-77.

El Messaoudi, N., El Khomri, M., Lacherai, A., Bentahar, S., Dbik, A., Bakiz, B., (2017). Valorization and characterization of wood of the jujube shell: Application to the removal of cationic dye from aqueous solution. *J. Eng. Sci. Technol.* 12(2), 423–438.

El Qada, E.N., Allen, S.J., Walker, G.M., (2008). Adsorption of basic dyes from aqueous solution onto activated carbons. *Chem. Eng. J.* 135(3), 174-184.

Elwakeel K.Z., El-Bindary A.A., El-Sonbati A. Z., Hawas, R.A., (2017). Magnetic alginate beads with high basic dye removal potential and excellent regeneration ability. *Can. J. Chem.* 95(8), 807-815.

Eren, E., (2009). Investigation of a basic dye removal from aqueous solution onto chemically modified Unye bentonite. *J. Hazard. Mater.* 166(1), 88–93.

Eskhan, A., Banat, F., Selvaraj, M., Haija, M.A., (2018a). *Grafting of poly (styrene-co-maleic anhydride) onto calcium alginate hydrogel for enhancing the adsorptive removal of methyl violet 6B cationic dye from aqueous solutions.* 2–3.

Eskhan, A., Banat, F., Selvaraj, M., Haija, M.A., (2018b). Enhanced removal of methyl violet 6B cationic dye from aqueous solutions using calcium alginate hydrogel grafted with poly (styrene-co-maleic anhydride). *Polym. Bull.* 1–29.

Essawy, A.A., Ali, A.E.H., Abdel-Mottaleb, M.S.A. (2008). Application of novel copolymer-TiO$_2$ membranes for some textile dyes adsorptive removal from aqueous solution and photocatalytic decolorization. *J. Hazard. Mater*, 157 (2-3), 547-552.

Forgacs, E., Cserhati, T., Oros, G., (2004). Removal of synthetic dyes from wastewater: A review. *Env. Int.* 30(7), 953-971.

Ghayeni, S.B., Beatson, P.J., Schneider, R.P., Fane, A.G., 1998. Water reclamation from municipal wastewater using combined microfiltration-reverse osmosis (ME-RO): Preliminary performance data and microbiological aspects of system operation. *Desalination.* 116(1), 65-80.

Gogate, P.R., Pandit, A.B., (2004). A review of imperative technologies for wastewater treatment I: Oxidation technologies at ambient conditions. *Adv. Environ. Res.* 8(3-4), 501–551.

Grassi, M., Kaykioglu, G., Belgiorno, V., Lofrano, G., (2012). Removal of emerging contaminants from water and wastewater by adsorption process: In *Emerging compounds removal from wastewater* 2012, 15-37.

Guibal, E., McCarrick, P., Tobin, J.M. (2003). Comparison of the sorption of anionic dyes on activated carbon and chitosan derivatives from dilute solutions, *Sep. Sci. Technol.* 38(12-13), 3049-3073.

Guibal, E., Von Offenberg Sweeney, N., Vincent, T., Tobin, J.M. (2002). Sulfur derivatives of chitosan for palladium sorption. *React. Funct. Polym.* 50(2), 149-163.

Gupta, A.K., Pal, A., Sahoo, C. (2006). Photocatalytic degradation of a mixture of Crystal Violet (Basic Violet 3) and Methyl Red dye in aqueous suspensions using Ag$^+$ doped TiO$_2$. *Dye. Pigment.* 69(3), 224–232.

Gupta, V.K. (2009). Application of low-cost adsorbents for dye removal-A review. *J. Environ. Manage.* 90(8), 2313-2342.

Hamutoğlu, R., Dinçsoy, A.B., Cansaran-Duman, D., Aras, S., (2012). Biyosorpsiyon, adsorpsiyon ve fitoremediasyon yöntemleri ve uygulamalarl. *Turk Hij. ve Deney. Biyol. Derg.* 69(2012), 69.

Hokkanen, S., Bhatnagar, A., Sillanpää, M., (2016). A review on modification methods to cellulose-based adsorbents to improve adsorption capacity. *Water Res.* 91, 156–173.

Jassal, P., Raut, V., (2015). Removal of crystal violet from wastewater using different chitosans and cross-linked derivatives, *WIT Transactions on Ecology and the Environment,* 196, 495-504. 196, 495–504.

Jayganesh, D., Tamilarasan, R., Kumar, M., Murugavelu, M., Sivakumar, V., (2017). Equilibrium and modelling studies for the removal of crystal violet dye from aqueous solution using eco-friendly activated carbon prepared from Sargassm wightii seaweeds. *J. Mater. Environ. Sci.* 8, 2122–2131.

Kanamadi, R.D., Ahalya, N., Ramachandra, T.V. (2006). Low cost biosorbents for dye removal. *CES Technical Report*, 113, 1-121.

Kannan et al. (2009). Removal of plant poisoning dyes by adsorption on tomato plant root and green carbon from aqueous solution and its recovery. *Desalination* 249(3), 1132–1138.

Kannan, N., Sundaram, M.M., 2001. Kinetics and mechanism of removal of methylene blue by adsorption on various carbons—a comparative study. *Dye. Pigment.* 51(1), 25–40.

Keyhanian, F., Shariati, S., Faraji, M., Hesabi, M., (2016). Magnetite nanoparticles with surface modification for removal of methyl violet from aqueous solutions. *Arab. J. Chem.* 9 (2016), S348–S354.

Khan, N.A., Hasan, Z., Jhung, S.H., (2013). Adsorptive removal of hazardous materials using metal-organic frameworks (MOFs): A review. *J. Hazard. Mater.* 244, 444-456.

Kulkarni, M.R., Revanth, T., Acharya, A., Bhat, P., (2017). Removal of Crystal Violet dye from aqueous solution using water hyacinth: Equilibrium, kinetics and thermodynamics study. *Resour. Technol.* 3(1), 71–77.

Kumar, R., Ahmad, R., (2011). Biosorption of hazardous crystal violet dye from aqueous solution onto treated ginger waste (TGW). *Desalination* 265(1-3), 112–118.

Lairini, S., El Mahtal, K., Miyah, Y., Tanji, K., Guissi, S., Boumchita, S., Zerrouq, F., (2017). The adsorption of Crystal violet from aqueous solution by using potato peels (*Solanum tuberosum*): Equilibrium and kinetic studies. *J. Mater. Environ. Sci.* 8(9), 3252–3261.

Laskar, N., Kumar, U., (2018). Adsorption of crystal violet from wastewater by modified bambusa tulda. *J. Civ. Eng.* 22(8), 2755–2763.

Laysandra, L., Santosa, F.H., Austen, V., Soetaredjo, F.E., Foe, K., Putro, J.N., Ju, Y.H., Ismadji, S., (2018). Rarasaponin-bentonite-activated biochar from durian shells composite for removal of crystal violet and Cr(VI) from aqueous solution. *Environ. Sci. Pollut. Res.* 25(30), 30680–30695.

Li, X., Zheng, H., Wang, Y., Sun, Y., Xu, B., Zhao, C., (2017). Fabricating an enhanced sterilization chitosan-based flocculants: Synthesis, characterization, evaluation of sterilization and flocculation. *Chem. Eng. J.* 319(2017), 119–130.

Lin, Y., He, X., Han, G., Tian, Q., Hu, W., (2011). Removal of crystal violet from aqueous solution using powdered mycelial biomass of *Ceriporia lacerata* P2. *J. Environ. Sci.* 23(12), 2055-2062.

Ling, S.L.Y., Yee, C.Y., Eng, H.S., (2011). Removal of a cationic dye using deacetylated chitin (chitosan). *J. Appl. Sci.* 11(8), 1445-1448.

Liu, B., Chen, X., Zheng, H., Wang, Y., Sun, Y., Zhao, C., Zhang, S., (2018). Rapid and efficient removal of heavy metal and cationic dye by carboxylate-rich magnetic chitosan flocculants: Role of ionic groups. *Carbohydr. Polym.* 181, 327–336.

Liu, B., Zheng, H., Wang, Y., Chen, X., Zhao, C., An, Y., Tang, X., (2018). A novel carboxyl-rich chitosan-based polymer and its application for clay flocculation and cationic dye removal. *Sci. Total Environ.* 640, 107–115.

Liu, J., Wang, Y., Zhang, X., Fang, Y., Mwamulima, T., Song, S., Peng, C., (2018). Preparation of Fe@GAC and Fe@GAR and their application for removal of crystal violet from wastewater. *Water. Air. Soil Pollut.* 229(2), 38.

Mahamadi, C., Mawere, E., (2013). Kinetic modeling of methylene blue and crystal violet dyes adsorption on alginate-fixed water hyacinth in single and binary systems. *Am. J. Anal. Chem.* 4(10), 17–24.

Mahdavinia, G.R., Aghaie, H., Sheykhloie, H., Vardini, M.T., Etemadi, H., (2013). Synthesis of CarAlg/MMt nanocomposite hydrogels and adsorption of cationic crystal violet. *Carbohydr. Polym.* 98(1), 358–365.

Mani, S., Bharagava, R.N., (2016). Exposure to crystal violet, its toxic, genotoxic and carcinogenic effects on environment and its degradation and detoxification for environmental safety, In *Rev Environ Contam Toxicol.* 237, 71-104.

Michaels G.B, Lewis D.L., (1986). Microbial transformation rates of azo and triphenylmethane dyes. *Env. Toxicol. Chem.* 5(2), 161–166.

Mittal, A., Mittal, J., Malviya, A., Kaur, D., Gupta, V.K., (2010). Adsorption of hazardous dye crystal violet from wastewater by waste materials. *J. Colloid Interface Sci.* 343(2), 463-473.

Miyah, Y., Lahrichi, A., Idrissi, M., Anis, K., Kachkoul, R., (2017). Removal of cationic dye, Crystal Violet in aqueous solution by the local clay, *J Mater Environ Sci,* 8(10), 3570-3582.

Mona, S., Kaushik, A., Kaushik, C.P., (2011). Waste biomass of *Nostoc linckia* as adsorbent of crystal violet dye: Optimization based on statistical model. *Int. Biodeterior. Biodegrad.* 65(3), 513-521.

Mori, M., Sekine, Y., Hara, N., Nakarai, K.-I., Suzuki, Y., Kuge, H., Kobayashi, Y., Arai, A., Itabashi, H., (2013). Adsorptivity of heavy metals Cu(II), Cd(II), and Pb(II) on woodchip-mixed porous mortar. *Chem. Eng. J.* 215, 202-208.

Muthukumaran, C., Sivakumar, V.M., Thirumarimurugan, M., (2016). Adsorption isotherms and kinetic studies of crystal violet dye removal from aqueous solution using surfactant modified magnetic nano-adsorbent. *J. Taiwan Inst. Chem. Eng.* 63, 354–362.

Nakhjiri, M.T., Marandi, G.B., Kurdtabar, M., (2018). Poly (AA-co-VPA) hydrogel cross-linked with N-maleyl chitosan as dye adsorbent:

Isotherms, kinetics and thermodynamic investigation. *Int. J. Biol. Macromol.* 117, 152–166.

Oladipo, A.A., Gazi, M., (2014). Enhanced removal of crystal violet by low cost alginate/acid activated bentonite composite beads: Optimization and modelling using non-linear regression technique. *J. Water Process Eng.* 2, 43–52.

Omer, O.S., Hussein, M.A., Hussein, B.H.M., Mgaidi, A., (2018). Adsorption thermodynamics of cationic dyes (methylene blue and crystal violet) to a natural clay mineral from aqueous solution between 293.15 and 323.15 K. *Arab. J. Chem.* 11(5), 615–623.

Pal, A., Pan, S., Saha, S., (2013). Synergistically improved adsorption of anionic surfactant and crystal violet on chitosan hydrogel beads. *Chem. Eng. J.* 217, 426–434.

Parshetti, G.K., Parshetti, S.G., Telke, A.A., Kalyani, D.C., Doong, R.A., Govindwar, S.P., (2011). Biodegradation of Crystal Violet by *Agrobacterium radiobacter*. *J. Environ. Sci.* 23(8), 1384–1393.

Patil, S., Deshmukh, V., Renukdas, S., Patel, N., Mahavidyalaya, Y., (2011). Kinetics of adsorption of crystal violet from aqueous solutions using different natural materials. *Int. J. Environ. Sci.* 1(6), 1116–1134.

Pillai, I.M.S., Gupta, A.K., Sahoo, C., (2011). Electrochemical oxidation of crystal violet dye (basic violet 3) using lead oxide electrodes. *Environ. Manag. Eng.* 60-66.

Puthiya Veetil Nidheesh, Rajan Gandhimathi, S.T.R. and T.S.A.S., (2018). Adsorption and desorption characteristics of crystal violet in bottom ash column, *J Urban Environ Eng* 6(1), 18-29.

Qin, J., Qiu, F., Rong, X., Yan, J., Zhao, H., Yang, D., (2015). Adsorption behavior of crystal violet from aqueous solutions with chitosan-graphite oxide modified polyurethane as an adsorbent. *J. Appl. Polym. Sci.* 132(17), 1–10.

Rehman, F., Sayed, M., Khan, J.A., Khan, H.M., (2017). Removal of crystal violet dye from aqueous solution by gamma irradiation. *J. Chil. Chem. Soc.* 62(1), 3359–3364.

Ren, X., Xiao, W., Zhang, R., Shang, Y., Han, R., (2015). Adsorption of crystal violet from aqueous solution by chemically modified phoenix tree leaves in batch mode. *Desalin. Water Treat.* 53(5), 1324–1334.

Ren, Y., Chen, Y., Sun, M., Peng, H., Huang, K., (2014). Rapid and efficient removal of cationic dyes by magnetic chitosan adsorbent modified with EDTA. *Sep. Sci. Technol.* 49(13), 2049–2059.

Robinson, T., Mcmullan, G., Marchant, R., Nigam, P., (2001). Remediation of dyes in textile effluent: a critical review on current treatment technologies with a proposed alternative. *Bioresour. Technol.* 77(3), 247-255.

Ruan, W., Hu, J., Qi, J., Hou, Y., Cao, R., Wei, X., (2018). Removal of crystal violet by using reduced-graphene-oxide-supported bimetallic Fe/Ni nanoparticles (rGO/Fe/Ni): Application of artificial intelligence modeling for the optimization process. *Materials.* 11(5), 865.

Saeed, A., Sharif, M., and Iqbal, M. (2010). Application potential of grapefruit peel as dye sorbent: Kinetics, equilibrium and mechanism of crystal violet adsorption. *J. Hazard. Mater.* 179(1-3), 564-572.

Saha, T.K., Karmaker, S., Ichikawa, H., Fukumori, Y., (2005). Mechanisms and kinetics of trisodium 2-hydroxy-1, 1'-azonaphthalene-3,4', 6-trisulfonate adsorption onto chitosan. *J. Colloid Interface Sci.* 286(2), 433-439.

Sakkayawong, N., Thiravetyan, P. and Nakbanpote, W. (2005). Adsorption mechanism of synthetic reactive dye wastewater by chitosan. *J. Colloid Interface Sci.* 286(1), 36-42.

Salleh, M.A.M., Mahmoud, D.K., Karim, W.A.W.A., Idris, A., (2011). Cationic and anionic dye adsorption by agricultural solid wastes: A comprehensive review. *Desalination.* 280(1-3), 1-13.

Sarkar, K., Debnath, M., Kundu, P.P., (2012). Hydrology recyclable crosslinked o-carboxymethyl chitosan for removal of cationic dye from aqueous solutions. *Hydrol Curr. Res* 3, 1-9.

Sarma, G.K., Sen Gupta, S., Bhattacharyya, K.G., (2016). Adsorption of Crystal violet on raw and acid-treated montmorillonite, K10, in aqueous suspension. *J. Environ. Manage.* 171, 1–10.

Shouman, M.A., Khedr, S.A., Attia, A.A., (2012). Basic dye adsorption on low cost biopolymer: Kinetic and equilibrium studies. *IOSR J. Appl. Chem.* 2, 27–36.

Kartina S.A.K, Lim, S-F., Chua D., Salleh S.F., Law P.L. (2018). Removal of crystal violet and acid green dye in aqueous solution using banana plant-derived sorbents, *Malaysian J. Anal. Sci.* 22(1), 115–122.

Skjak-Braek, G., Grasdalen, H., (1989). Inhomogeneous polysaccharide ionic gels. *Carbohydr. Polym.* 10(1), 31–54.

Smitha, T., Thirumalisamy, S. and Manonmani, S. (2012). Equilibrium and kinetics study of adsorption of crystal violet onto the peel of *Cucumis sativa* fruit from aqueous solution. *Journal Chem.* 9(3), 1091–1101.

Su, C., Wang, Y., (2011). *Influence factors and kinetics on crystal violet degradation by Fenton and optimization parameters using response surface methodology* 15, 76–80.

Sun, P., Hui, C., Azim Khan, R., Du, J., Zhang, Q., Zhao, Y.-H., (2015). Efficient removal of crystal violet using Fe_3O_4-coated biochar :The role of the Fe_3O_4 nanoparticles and modeling study their adsorption behavior. *Sci. Rep.* 5, 12638.

Sun, P., Hui, C., Wang, S., Wan, L., Zhang, X., Zhao, Y., (2016). *Bacillus amyloliquefaciens* biofilm as a novel biosorbent for the removal of crystal violet from solution. *Colloids Surf. B: Biointerfaces* 139, 164–170.

Tahir, N., Bhatti, H.N., Iqbal, M., Noreen, S., (2017). Biopolymers composites with peanut hull waste biomass and application for crystal violet adsorption. *Int. J. Biol. Macromol.* 94, 210–220.

Vardikar, H.S., Bhanvase, B.A., Rathod, A.P., Sonawane, S.H., (2018). Sonochemical synthesis, characterization and sorption study of Kaolin-Chitosan-TiO_2 ternary nanocomposite: Advantage over conventional method. *Mater. Chem. Phys.* 217, 457–467.

Slokar, Y.M. and Le Marechal, A.M., (1998). Methods of decoloration of textile wastewaters. *Dye. Pigment.* 37(4), 335-356.

Watters J.C., Biagtan E, Senler, O., 1991. Ultrafitration of a textile plant efflent. *Sep. Sci. Technol.* 26(10-11), 1295–1313.

Wong, Y.C., Szeto, Y.S., Cheung, W.H., McKay, G., (2008). Effect of temperature, particle size and percentage deacetylation on the adsorption of acid dyes on chitosan. *Adsorption.* 14(1), 11-20.

Zhou, Y., Zhang, M., Wang, X., Huang, Q., Min, Y., Ma, T., Niu, J., (2014). Removal of crystal violet by a novel cellulose-based adsorbent: Comparison with native cellulose. *Ind. Eng. Chem. Res.* 53(13), 5498–5506.

In: Crystal Violet
Editor: Victor Duffet
ISBN: 978-1-53615-806-9
© 2019 Nova Science Publishers, Inc.

Chapter 3

REMOVAL OF CRYSTAL VIOLET DYE FROM AQUEOUS SOLUTION USING ASH-BASED ADSORBENT MATERIALS

Denise A. Fungaro*, Suzimara Rovani, Tharcila C. R. Bertolini and Flamarion F. Filho

Instituto de Pesquisas Energéticas e Nucleares,
IPEN–CNEN/SP, São Paulo, SP, Brazil

ABSTRACT

Crystal Violet (CV) is widely used for various purposes and enters into the aquatic systems from the effluents of textile, paint, medical and biotechnological industries. A considerable amount of this dye is lost during manufacturing and processing operations. Contaminated wastewater containing CV must be treated before releasing in the environment because it is highly cytotoxic and carcinogenic to mammalian cells, present mitotic poisoning nature and is non-biodegradable being classified as a recalcitrant molecule. This chapter reports the removal of CV dye from water using surfactant-modified zeolite from coal fly ash (MZSF), surfactant-modified zeolite from coal bottom ash (MZSB) and nanosilica from sugarcane waste ash (SiO_2NP).

---

* Corresponding Author's E-mail: dfungaro@ipen.br.

The adsorbent materials were characterized to obtain chemical and mineralogical composition and others physicochemical properties. The adsorption kinetic of CV onto adsorbents was discussed using the pseudo-first order, pseudo-second order, and Elovich models. The Langmuir and Freundlich isotherm models were used to describe the equilibrium adsorption data. The maximum adsorption capacities were 36.7 mg g^{-1} and 21.1 mg g^{-1} for CV/MZSF and CV/MZSB, respectively. The adsorption process of CV/SiO$_2$NP achieves equilibrium in 60 min of contact time, and the maximum adsorption capacity was 117.98 mg g^{-1}. Application of the adsorbent materials synthesized from agricultural waste and coal combustion products can ensure the sustainability and cost-effectiveness of treating effluent containing CV dye, especially effluent from the textile industries generated in large quantity.

Keywords: crystal violet, zeolite, coal ashes, nanosilica

1. INTRODUCTION

Dyes are common contaminants in the effluents of textile industries, leather tanning, cosmetics, paper, food processing and pharmaceuticals (Bertolini et al. 2013). They are classified as anionic, cationic and nonionic. The dye anionic include acid, reactive, azo and direct dyes; cationic include basic dyes; and nonionic include dispersed dyes which do not ionize in the aqueous medium (Salleh et al. 2011; Sharma & Kaur 2018). Among these classes of dyes, the cationic have higher toxicity than anionic because they can interact with the surface of negatively charged membranes and can enter the cells and concentrated in the cytoplasm (Rehman et al. 2017).

Some of the basic dyes are methylene blue, crystal violet and these usually are used, in the textile industry, in the dyeing of acrylic, wool, nylon, and silk (Salleh et al. 2011). Due to their complex aromatic structures, they are generally non-biodegradable, exhibit high stability and have high toxicity with carcinogenic properties (Hamza et al. 2018; Hung 2018). Among the cationic dyes, the crystal violet (Figure 1) is the most toxic; it is also known as gentian violet or methyl violet 10B (Gusmão et al. 2013).

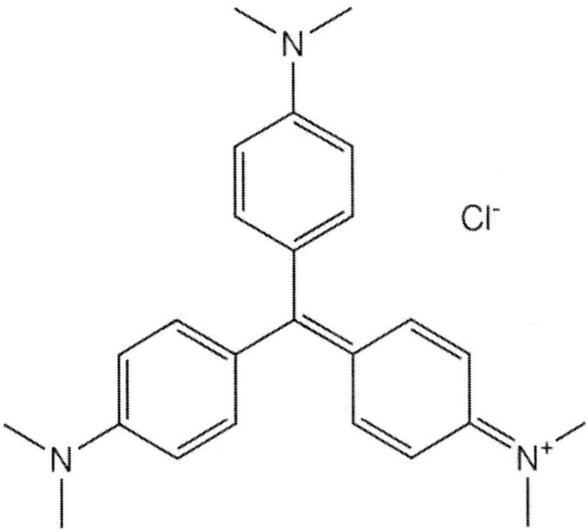

Figure 1. Chemical structure of crystal violet dye (MM = 408 g mol^{-1}).

The original crystal violet dye production procedure was developed in 1883 (Caro & Kern 1883) and details of this and others production procedures are described in literature (Mani & Bharagava 2016). Crystal violet is also a pH indicator used in the range of 0 to 1.8 when it turns from yellow to blue. Its molecular formula is $C_{25}H_{30}N_3{}^+Cl^-$, its maximum wavelength ($\lambda_{máx}$) is at 590 nm with an extinction coefficient (ε) of 87,000 M^{-1} cm^{-1}, and IUPAC name is Tris (4-(dimethylamino) phenyl) methylium chloride (Mani & Bharagava 2016).

Crystal violet is used for different purposes, such as bactericidal agent; bacterial analysis by Gram staining; fabric dyeing; printing paper; veterinary medicine as an additive in animal feed, mainly poultry; skin marking before plastic surgeries; cure stomatitis and canker sores (Chowdhury et al. 2013; Kulkarni et al. 2017; Rehman et al. 2017).

Although crystal violet is used in human and veterinary medicine, frequent use can cause serious health effects, because this dye is known for its mutagenic, teratogenic and mitotic poison nature (Kulkarni et al. 2017). Crystal violet can cause moderate eye irritation and even permanent damage to the cornea. It is highly toxic to mammalian cells, under extreme conditions it can lead to respiratory and renal failure. This dye can also cause bladder cancer in humans and cancer of the digestive system in other animals. More effects of toxic, genotoxic and carcinogenic of crystal violet in the environment are published in detail in the book chapter (Mani & Bharagava 2016).

Treatment of effluents containing dyes, like crystal violet, is not so simple. They are recalcitrant species, that is, they are organic compounds that are hardly degraded, are hydrophobic, bioaccumulative and have high stability (Kulkarni et al. 2017). Several techniques have been applied to the removal of dyes. Some of these techniques are adsorption; flocculation combined with flotation; coagulation; membrane filtration; electrokinetics; ozonation; oxidation; precipitation; ion exchange (Salleh et al. 2011).

According to the literature, one of the most widely used methods for the removal of crystal violet from aqueous effluents is the adsorption due to their simplicity, high efficiency, and low cost, as well as the availability of a wide range of adsorbents that can be applied (Kulkarni et al. 2017). This process transfers the staining from the aqueous effluent to a solid phase (adsorbent), significantly decreasing the bioavailability of the dye to living organisms. The decontaminated effluent can then be released into the environment, or water can be reused in the industrial process. Subsequently, the adsorbent can be stored in a dry place, without direct contact with the environment.

A large number of works on adsorption of crystal violet using solid adsorbents were published in the last years. Table 1 shows different maximum adsorption capacities of crystal violet obtained with different adsorbents used in the adsorption process.

The adsorbent materials studied by our group are zeolites from coal fly and bottom ashes (Fungaro et al. 2009, 2010, 2013; Carvalho et al. 2011; Izidoro et al. 2012, 2013, 2018; Bertolini et al. 2013, 2015; Magdalena et al. 2014; Alcantara et al. 2015, 2016; Cunico et al. 2015, 2018) and silica from sugarcane waste ash (Alves et al. 2017; Rovani et al. 2018a, 2018b).

All structure-crystalline substance characterized by a framework of interconnected tetrahedra, each consisting of four oxygen atoms involving a cation, are known as zeolites. They were considered hydrated aluminosilicates of alkali metals. Nowadays conceptualization of zeolites has expanded, admitting other cations that not only Si and Al, but these still prevail (Resende et al. 2008). Coal fly and bottom ashes can be converted into zeolites, using a simple method, because these wastes are rich in Si and Al.

Table 1. Adsorbents used for crystal violet removal by adsorption process

Adsorbents	$Q^0_{max.}$ (mg g⁻¹)	References
Moroccan pyrophyllite	12.19	(Miyah et al. 2017)
Modified bambusa tulda	20.84	(Laskar & Kumar 2018)
Coniferous pinus bark powder	28.24	(Ahmad 2009)
Nanosilica-supported poly β-cyclodextrin	34.5	(Chen et al. 2018)
Nanosilica from rice husk	42.0	(Hung 2018)
Biofibers	54.0	(Andreou & Pashalidis 2018)
Eggshells	70.03	(Chowdhury et al. 2013)
Palm kernel fiber	78.9	(El-Sayed 2011)
Tunisian smectite clay	86.54	(Hamza et al. 2018)
Chitosan aniline composite	100.6	(Tahir et al. 2017)
Activated bentonite	108.56	(Bellir et al. 2012)
Sugarcane bagasse modified with EDTA dianhydride	327.87	(Gusmão et al. 2013)
Natural clay mineral	330.0	(Omer et al. 2018)
Nanocomposite (polyacrylamide grafted xanthan gum/nanosilica)	378.8	(Ghorai et al. 2014)
Bio-nanosilica obtained from rice husk	495.0	(Peres et al. 2018)
Bentonite – alginate composite	601.93	(Fabryanty et al. 2017)
Activated carbon from *Ferula orientalis*	769.23	(Aysu & Küçük 2015)

Silica is a combination of silicon and oxygen in the form of SiO_2, and, all the minerals whose only cation is silicon, can be called silica. However, there are two known forms of SiO_2 the amorphous and crystalline form (Foletto et al. 2005).

According to the literature, rice husk and sugarcane bagasse are rich in Si. Some researchers have developed methods to extract the amorphous silica of such wastes in the most economical way and as much as possible through a chemical and/or thermal treatment (Kalapathy et al. 2000a, 2002,

Alves et al. 2017). The silica is extracted in the pure form by solubilization under alkaline conditions and subsequent precipitation in a lower pH (Iler 1979; Kamath & Proctor 1998; Kalapathy et al. 2000b).

Considering these aspects, this chapter describes the application of zeolite from coal ashes and nanosilica from sugarcane waste ash to remove the crystal violet from aqueous solution using the adsorption method.

2. EXPERIMENTAL SECTION

2.1. Materials

The crystal violet dye (CV; CI 42555; $C_{25}H_{30}N_3^+Cl^-$; 408 g mol^{-1}) was obtained from Proton-Research and considered as purity 100%. Coal fly (CFA) and bottom ashes (CBA) samples were donated by Jorge Lacerda Thermoelectric Complex (Santa Catarina, Brazil). Sugarcane waste ash sample was donated by COSAN S.A. (São Paulo, Brazil). The ashes CFA and CBA were used without any pretreatment. The sugarcane waste ash was pre-treatment according to the reference (Kalapathy et al. 2000). Sodium hydroxide (97%), and sodium aluminate (100%) obtained from Sigma-Aldrich Pty. Ltd. (Australia) were used in zeolite synthesis. The surfactant hexadecyltrimethylammonium bromide (HDTMA-Br) with molar mass 364.46 g mol^{-1} obtained from Merck (Germany) was used for surfactant-modified zeolite synthesis. Sodium hydroxide micro pearls (> 99%), and hydrochloric acid (35 - 37%) obtained from Synth (Brazil) were used in nanosilica synthesis.

2.2. Synthesis of Adsorbents

2.2.1. Synthesis of Surfactant-Modified Zeolite

The zeolite was synthesized by hydrothermal treatment. The coal fly or bottom ash (20 g) was mixed with 160 mL of 3.5 mol L^{-1} aqueous NaOH solution in a Teflon vessel. This mixture was heated to 100°C in the oven for 24 h (Henmi 1987). The zeolitic material was repeatedly washed with deionized water to remove excess sodium hydroxide until the washing water the pH~10; then it was dried at 50°C for 12 h (Bertolini et al. 2013). The modified zeolite was prepared by mixing of 10 g of zeolite from coal fly ash or bottom ash with 0.2 L of 1.8 mmol L^{-1} HDTMA-Br. The mixture of zeolite and HDTMA-Br solution was stirred for 7 h at 120 rpm and 25°C. The suspension was filtered, and the solid was dried in the oven at 50°C for 12 h. The modified zeolites were labeled as MZSF and MZSB for zeolite prepared with fly ash and bottom ash, respectively (Bertolini et al. 2015; Fungaro et al. 2016).

2.2.2. Synthesis of Silica Nanoparticles

Extraction of sodium silicate from sugarcane waste ash (SWA) was carried out by reaction of sodium hydroxide melted and SWA at 400°C for 1 h, the proportion (w:w) of ash:NaOH (1:1.5) (Rafiee et al. 2012; Hassan et al. 2014; Alves et al. 2017; Rovani et al. 2018b). Subsequently, it was added to the mixture distilled water and refluxed (boiling temperature) for 4 h to leave all the sodium silicate dissolved in aqueous medium. Then, hydrochloric acid solution (6.0 mol L^{-1}) was added, dropwise, a until pH decrease to 2.0 (Rafiee et al. 2012; Hassan et al. 2014; Rovani et al. 2018a). This nanosilica gel was washed with distilled water, filtered and oven dried at 120 °C overnight.

2.3. Characterization of Materials

The materials were characterized by XRF, XRD, FTIR, and SEM, whose details are in the references (Izidoro et al. 2012; Bertolini et al. 2013, 2015; Fungaro et al. 2016; Alves et al. 2017; Rovani et al. 2018a, 2018b).

The determination of the pH value at which the nanosilica surface is electrically neutral (point of zero charge - PZC) was performed based on the batch system equilibrium method. The point of zero charge of the nanosilica was performed by adding 40 mL of the solution of the electrolyte 0.10 mol L^{-1} NaNO$_3$ to 0.20 g of nanosilica. The pH values were adjusted between 2.0 and 11.0 by the addition of 0.10 mol L^{-1} NaOH solution or HNO$_3$. Suspensions were shaken for 24 h at 25°C and 190 rpm. The pH was determined before (initial pH (pH$_i$)) and after the contact time (final pH (pH$_f$)) of 24 h. The value of pH$_{PZC}$ is the point at which the curve of ΔpH (pH$_f$ - pH$_i$) versus pH$_i$ crosses the line of the abscissa y at zero (Mahmood et al. 2011). The pH$_{PZC}$ of surfactant-modified zeolites was determinate previously in reference (Bertolini et al. 2015; Fungaro et al. 2016).

2.4. Adsorption Studies

The crystal violet dye (CV) removal tests in water using surfactant-modified zeolite from coal fly ash (MZSF), surfactant-modified zeolite from bottom ash (MZSB), and silica nanoparticles (SiO$_2$NP) were performed in batch.

Aliquots of crystal violet solution with known concentrations were placed in beakers with adsorbent dose: 10 g L^{-1} of MZSF or 10 g L^{-1} of MZSB or 1 g L^{-1} of SiO$_2$NP. The suspensions were shaken on a shaker table (Quimis, Brazil) at 120 rpm for MZSF and MZSB, and at 190 rpm for SiO$_2$NP in different time intervals (Table 2). All experiments were performed in triplicate and at 25°C ± 2°C. Analytical curve prepared for crystal violet is presented in Figure 2, the concentration of crystal violet before and after adsorption was detected by a UV-visible spectrophotometer (Cary 1E, Varian) in the maximum wavelength ($\lambda_{máx}$) of 590 nm.

For the adsorption isotherms study, the aqueous solution of crystal violet was placed in beakers with the same adsorbent dose above describe. The concentration was range from 23.4 to 579 mg L^{-1} and from 5 to 130 mg L^{-1} for MZSF or MZSB, and for SiO$_2$NP, respectively.

The equation of the analytical curve (insert in Figure 2) was A = - 0.01706 + 0.17233 [crystal violet dye] (mg L^{-1}), and the value of the coefficient of determination adjusted (R$^2_{adj.}$) was 0.99445. The amount of dye removal was expressed as the removal percentage of crystal violet and calculated by the equation 1:

Table 2. Concentration of CV and time interval in the adsorption kinetics study

Adsorbents	Initial concentration (mg L⁻¹)	Contact time (min)
MZSF	185	1 - 12
MZSB		2 - 20
SiO$_2$NP	20 and 50	0 - 360

Source: Adapted from reference (Bertolini et al. 2015).

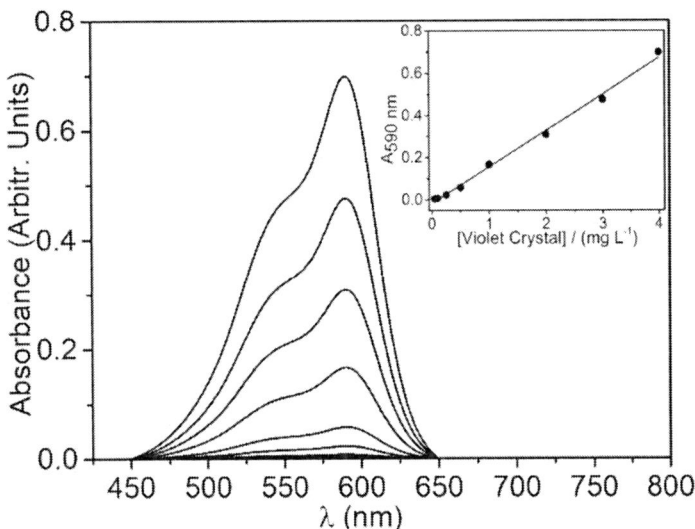

Figure 2. UV-visible spectra of solution containing crystal violet dye at concentrations from 0.05 to 4.0 mg L⁻¹. Insert: analytical curve of crystal violet dye.

$$\% \, Removal = \frac{(C_i - C_f)}{C_i} \times 100 \quad \text{(Eq. 1)}$$

where C_i and C_f are the initial and final concentration of crystal violet, respectively. The amount of dye adsorption as a function of time and at equilibrium, q_t, and q_e (mg g⁻¹), respectively, were calculated using the following equations 2 and 3:

$$q_e = \frac{(C_i - C_e)}{m} \times V \qquad \text{(Eq. 2)}$$

$$q_t = \frac{(C_i - C_f)}{m} \times V \qquad \text{(Eq. 3)}$$

where C_i, C_f, and C_e (mg L^{-1}) are concentrations of dye at initial, final and equilibrium, respectively, V (L) is the volume of crystal violet solution and m (g) is the mass of the adsorbent.

2.5. Linear and Non-Linear Kinetic Models

This study involves the relationship between adsorption efficiency and adsorbent/adsorbate contact time. Linear kinetic adsorption models used were pseudo-first order (Eq. 4) (Lagergren 1898; Kumar 2006), pseudo-second order (Eq. 5) (Blanchard et al. 1984; Kumar 2006), and Elovich (Eq. 6) (Mclintock 1967; Chien & Clayton 1980; Tran et al. 2017). These models were applied only for crystal violet removal by MZSF and MZSB.

Non-linear kinetic adsorption models used were pseudo-first order (Eq. 7) (Lagergren 1898; Ho & McKay 1998), pseudo-second order (Eq. 8) (Blanchard et al. 1984; Ho et al. 1996) and were applied only for crystal violet removal by SiO$_2$NP.

$$\log(q_e - q_t) = \log(q_e) - \frac{k_1 \times t}{2.303} \qquad \text{(Eq. 4)}$$

$$\frac{1}{q_t} = \left(\frac{1}{k_2 \times q_e^2}\right) \times \frac{1}{t} + \frac{1}{q_e} \qquad \text{(Eq. 5)}$$

$$q_t = \frac{1}{\beta}\ln(t) + \frac{1}{\beta}\ln(\alpha\beta) \qquad \text{(Eq. 6)}$$

$$q_t = q_e \cdot \left(1 - e^{(-k_1 t)}\right) \qquad \text{(Eq. 7)}$$

$$q_t = \frac{k_2 \cdot q_e^2 \, t}{1 + k_2 \cdot q_e \cdot t} \qquad \text{(Eq. 8)}$$

where q_t is the amount of adsorbate adsorbed at time t (mg g^{-1}), q_e is the equilibrium adsorption capacity (mg g^{-1}), k_1 is the pseudo-first-order rate constant (min^{-1}), and t is the contact time (min), k_2 is the pseudo-second-order rate constant (g mg^{-1} min^{-1}) (Lima et al. 2015). Elovich equation assumed $\alpha\beta t \gg 1$ (Chien & Clayton 1980) and a plot of qt versus ln$t$ should give a linear relationship with a slope of $(1/\beta)$ and an intercept of $(1/\beta)\ln(\alpha\beta)$.

The α (mg g^{-1} x min) is the initial rate constant because $dq_t/dt \to \alpha$ when $q_t \to 0$; and β (mg g^{-1}) is the desorption constant during any one experiment (Tran et al. 2017).

2.6. Equilibrium Models

The adsorption isotherms parameters were evaluated using the linear and non-linear regression methods of analysis. The equilibrium adsorption models utilized were Langmuir Hanes-Woolf linearization (Eq. 9); non-linear method (Eq. 10) (Langmuir 1918; Ho 2004; Tran et al. 2017), and Freundlich linear method (Eq. 11); non-linear method (Eq. 12) (Freundlich 1906; Tran et al. 2017). Linear models were applied only for crystal violet removal by MZSF and MZSB. The Chi-square (χ^2) test was employed, the lowest values were used to validate the applicability of isotherms tested (Eq. 13) (Ho 2004; Bertolini et al. 2015).

$$\frac{C_e}{q_e} = \left(\frac{1}{Q_{max}^0}\right) C_e + \left(\frac{1}{Q_{max}^0 \cdot K_L}\right) \quad \text{(Eq. 9)}$$

$$q_e = \frac{Q_{max}^0 \cdot K_L \cdot C_e}{1 + K_L \cdot C_e} \quad \text{(Eq. 10)}$$

$$\log q_e = \frac{1}{n \log C_e} + \log K_F \quad \text{(Eq. 11)}$$

$$q_e = K_F \cdot C_e^{\frac{1}{n_F}} \quad \text{(Eq. 12)}$$

$$X^2 = \Sigma \frac{(q_{e\,exp.} - q_{e\,cal.})^2}{q_{e\,calc.}} \quad \text{(Eq. 13)}$$

where q_e is the amount of adsorbate adsorbed at the equilibrium (mg g^{-1}), C_e is the adsorbate concentration at the equilibrium (mg L^{-1}). K_L is the Langmuir equilibrium constant (L mg^{-1}), Q_{max}^0 is the maximum adsorption capacity of the adsorbent (mg g^{-1}) assuming the formation of a monolayer of adsorbate over adsorbent. K_F is the Freundlich equilibrium constant [(mg g^{-1}) (L mg^{-1})$^{1/n}$], n_F is the Freundlich exponent (dimensionless) (Lima et al. 2015; Tran et al. 2017).

3. CRYSTAL VIOLET REMOVAL BY SURFACTANT-MODIFIED ZEOLITES

This part of chapter present an overview of the strategy used for the effective removal of Crystal Violet dye from aqueous solution using surfactant-modified zeolites from coal ashes described in the literature (Bertolini et al. 2015; Fungaro et al. 2016).

3.1. Characterizations of Zeolitic Materials

The physicochemical characterizations of the fly and bottom ashes of unmodified zeolites, and the surfactant-modified zeolites have been described in detail in the previous paper (Izidoro et al. 2012; Bertolini et al. 2013, 2015). Some important physicochemical characteristics of the surfactant-modified zeolites are presented briefly in Table 3.

The main constituents found for the samples of the modified zeolites are SiO_2, Al_2O_3, and Fe_2O_3 (Table 3). The zeolitic materials also exhibited a significant amount of Na element when compared with fly and bottom ashes due to hydrothermal treatment with NaOH solution. The bromide detected in the modified zeolites is the counterion present in the cation HDTMA adsorbed on the zeolite surface. The bulk density values found for MZSF and MZSB were 2.38 and 2.49 g cm^{-3}, respectively and are very close to the value found for the unmodified zeolites (Bertolini et al. 2013).

Table 3. Physicochemical properties of the surfactant-modified zeolites

Properties	MZSF	MZSB
Al$_2$O$_3$ (wt.%)	37	38
SiO$_2$ (wt.%)	36	33
Na$_2$O (wt.%)	8.7	8.0
Fe$_2$O$_3$ (wt.%)	8.5	13
CaO (wt.%)	3.4	2.7
TiO$_2$ (wt.%)	2.7	2.0
MgO (wt.%)	1.6	1.4
SO$_3$ (wt.%)	1.0	0.24
K$_2$O (wt.%)	0.8	0.7
ZnO (wt.%)	0.09	0.02
Br (wt.%)	0.06	0.05
ZrO$_2$ (wt.%)	0.06	0.04
Others (wt.%)	0.36	0.30
Bulk density (g cm^{-3})	2.38	2.49
BET surface area (m^2 g^{-1})	64	62
pH in water	8.3	8.5
[a]pH$_{PZC}$	6.3	6.1
[b]CEC (meq g^{-1})	1.61	1.49
Degree of hydrophobicity (%)	86	98

[a]point of zero charge; [b]cation exchange capacity.
Source: Adapted from references (Bertolini et al. 2015; Fungaro et al. 2016).

It was observed that the values of the specific surface area of the modified zeolites were higher than that of the precursor ash. This increase is due to the crystallization stage of the smooth spherical particles of the ashes after the hydrothermal treatment. The CEC value of the modified zeolite material of the fly ash was higher than the sample synthesized from the bottom ash. This is because fly ash has a smaller particle size (Bertolini et al. 2015; Fungaro et al. 2016).

The zeolites MZSF and MZSB presented (Table 3) point zero charge (pH$_{PZC}$) values lower than the pH in water evidencing that these materials present negative charge in aqueous solution (pH > pH$_{PZC}$). The change about the pH$_{PZC}$ is related to the formations of monolayer or bilayer structure on the zeolite surface (Li & Bowman 1998). The negative charge of MZSF and MZSB indicated the incomplete formation of a bilayer of surfactant on the surface of the materials (Bertolini et al. 2015; Fungaro et al. 2016).

The values of the degree of hydrophobicity of the modified zeolites presented on Table 3 were high and indicated that the surfactant was effectively adsorbed onto the material surface. Therefore, MZSF and MZSB zeolites preferentially interact with organic compound than with water (Bertolini et al. 2015).

The identification and interpretation of the crystalline phases present in the materials are prepared by comparing the diffraction database provided by "International Centre for Diffraction Data/Joint Committee on Powder Diffraction Standards" (ICDD/JCPDS). The XRD analyses of modified zeolites showed the crystalline phases were hydroxysodalite (JCPDS 31-1271) and NaX (JCPDS 38-0237) with peaks of quartz (JCPDS 85-0796) and mullite (JCPDS 74-4143) of ashes that remained after the synthesis (Bertolini et al. 2015; Fungaro et al. 2016).

Surfactant-modified zeolites presented structural parameters similar to those of the corresponding unmodified zeolites, evidencing that the crystalline nature remained intact after the adsorption of the surfactant molecules (Fungaro et al. 2013).

3.2. Adsorption Kinetic Studies

To study the kinetics of adsorption of CV, adsorption experiments were carried out at different contact time and a constant dye initial concentration and adsorbent dose. These experiments were performed against time (range 1 - 12 min for MZSF and 1 - 20 min for MZSB) and the effect of equilibrium time on the adsorption of CV onto the modified zeolites is presented in Figure 3.

These adsorption kinetic studies of the modified zeolites with CV have been described in the previous paper (Bertolini et al. 2015; Fungaro et al. 2016).

The dye removal was increased as the agitation time increased until equilibrium. It should be pointed out that apparent equilibrium was reached after 6 min for MZSF e 16 min para MZSB. The uptakes of CV for an initial concentration of 185 mg L^{-1} were found to be 77% and 52% for MZSF and MZSB, respectively (Bertolini et al. 2015).

The kinetics of adsorption of an adsorbate by any adsorbent is required for selecting optimum operating conditions for the pilot-scale batch process.

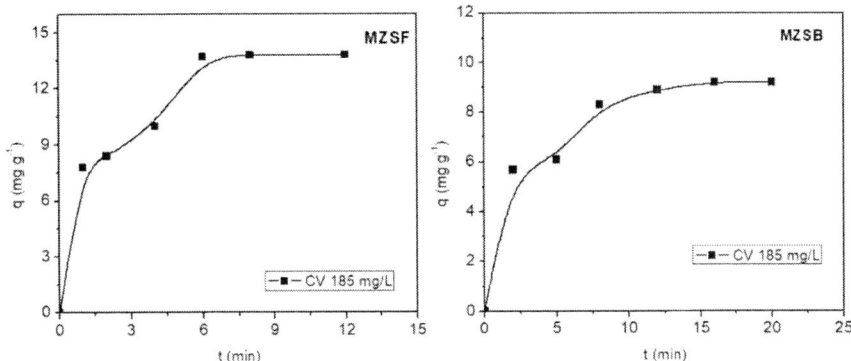

Figure 3. Influence of equilibrium time on the adsorption of CV on MZSF and MZSB. Source: Adapted from reference (Bertolini et al. 2015).

Aiming at evaluating the adsorption kinetics of CV onto modified zeolites, the pseudo-first order, pseudo-second order, and Elovich kinetic models were used to fit the experimental data. The values of kinetic constants for CV adsorption onto modified zeolites of the three kinetics equations along with correlation coefficients, R values, are given in Table 4.

The comparison of the three selected kinetic models showed that pseudo-second order model with the highest correlation coefficient values best describe the adsorption process for all system studied. Moreover, the generated experimental (q_e exp.) values of adsorption capacity were much closer to the theoretical values (q_e calc.). It was observed (Table 4) that the k_2 values are similar for both adsorbents. Thus, the adsorption rate was not influenced by the type of raw material of the adsorbents.

The kinetic parameters obtained from the Elovich model fitted to experimental data are also presented in Table 4. Elovich model is used for the chemisorption process and is suitable for systems with the heterogeneous adsorbing surface (Ho 2006). The parameter β is related to the extent of surface coverage; the values found for both modified zeolites are on the same order of magnitude. It was possible to observe that the initial adsorption rate (α) followed the order MZSF > MZSB, indicating that adsorption on modified zeolite from fly ash was faster than modified zeolite from bottom ash (Bertolini et al. 2015).

3.3. Adsorption Isotherm

The isotherms data of the modified zeolites were analyzed using two of the most commonly used equilibrium models namely the Langmuir and Freundlich isotherms. The Langmuir isotherm model suggests that there is no lateral interaction between the sorbed molecules and is based on monolayer coverage prediction of the adsorbate (Zou et al. 2011). The model of the Freundlich isotherm is based on multilayer adsorption on the heterogeneous surface (Langmuir 1918). The equilibrium studies of the modified zeolites with CV have been described in the previous paper (Bertolini et al. 2015; Fungaro et al. 2016).

The linear regressive method of least squares is used for finding the parameters of the isotherms. The relative parameters were obtained according to the intercept and slope from the linear plots (Bertolini et al. 2015). The nonlinear regressive method of least sum squares of the difference between calculated data and experimental data was used to determine the isotherm parameters with OriginPro 8.0. The isotherm parameters from the linear method and the nonlinear method were all listed in Table 5, respectively.

According to the results in Table 5, the CV/MZSF system presented the greater maximum adsorption capacity (Q_{max}^0). This can be attributed to higher loading amount of surfactant molecules on the external surface of zeolite from fly ash than zeolite from bottom ash due to their smaller particle size.

Table 4. Kinetic parameters for crystal violet adsorption onto MZSF and MZSB

	Adsorbents	
	MZSF	MZSB
Pseudo first-order		
k_1 (min^{-1})	7.78 x 10^{-1}	2.63 x 10^{-1}
$q_{e\,calc.}$ (mg g^{-1})	23.5	7.71
$q_{e\,exp.}$ (mg g^{-1})	13.8	9.20
R_1	0.878	0.969
Pseudo second-order		
k_2 (g mg^{-1} min^{-1})	4.37 x 10^{-2}	4.25 x 10^{-2}
h (mg g^{-1} min^{-1})	10.8	4.56
$q_{e\,calc.}$ (mg g^{-1})	15.7	10.4
$q_{e\,exp.}$ (mg g^{-1})	13.8	9.20
R_2	0.991	0.995
Elovich		
α (mg g^{-1} min^{-1})	34.5	18.9
β (g mg^{-1})	3.45 x 10^{-1}	5.67 x 10^{-1}
R_E	0.938	0.946

Source: Adapted from references (Bertolini et al. 2015; Fungaro et al. 2016).

Table 5. Langmuir and Freundlich isotherm parameters for CV adsorption on the adsorbents - comparison of linear and non-linear methods

Adsorbents	Langmuir		Freundlich	
	$Q^0_{max.}$ (mg g^{-1})	$^*b$	K_F [(mg g^{-1}) (L mg^{-1})$^{1/n}$]	n_F
Linear method				
MZSF	36.7	0.0413	4.03	2.36
MZSB	21.1	0.0116	1.07	2.07
Non-linear method				
MZSF	34.1	0.0479	5.29	2.81
MZSB	22.9	0.00842	1.14	2.12

$^*b = K_L \times q_e$

Source: Adapted from references (Bertolini et al. 2015; Fungaro et al. 2016).

The zeolite from coal fly ash will have particles sizes smaller than the material from coal bottom ash due to granulometric characteristics of the ash sample that served as raw material. Thus, when the particle size of a material decreases, it increases the external surface area, meaning an increase in the number of active sites available for adsorption (Ali & El-Bishtawi 1997; Bertolini et al. 2015).

The adsorption capacities obtained for unmodified zeolites from coal fly ash and bottom ash in the previous study were 19.6 and 17.6 mg g^{-1}, respectively. Thus, the modified zeolites synthesized in this study were more efficient in CV adsorption. This is due to the partition mechanism, which has been described in detail in the previous paper (Bertolini et al. 2013, 2015).

The values obtained of the test Chi-square (χ^2) and the correlation coefficients for each isotherm of the systems CV/MZSF and CV/MZSB obtained from linear and non-linear methods are presented in Table 6. When the two models are compared, from the linear regression, is observed the largest correlation coefficient value (R) and the lowest values for χ^2. For comparison of the models from the non-linear regression are considered the lowest values of deviation estimate (test Chi-square (χ^2)).

Table 6. Isotherm error deviation data related to the adsorption of CV onto the adsorbents

Adsorbents	Langmuir		Freundlich	
	R	χ^2	R	χ^2
Linear method				
MZSF	0.969	16.9	0.954	5.89
MZSB	0.946	4.06	0.976	1.72
Non-linear method				
MZSF	-	14.5	-	4.41
MZSB	-	5.96	-	1.64

Source: Adapted from reference (Bertolini et al. 2015).

As can be seen, the Langmuir model achieve correlation coefficient values near to unit, however lower values for χ^2 were observed in Freundlich equilibrium model both linear fits, as in the nonlinear adjustment.

Therefore, it was confirmed statistically that the Freundlich model was the best fit to the experimental data to describe the adsorption process of CV onto modified zeolites presented in this chapter.

Mohan et al. 2002 evaluated the removal of the crystal violet in the effluent. Coal fly ash was used as a low-cost adsorbent. The adsorption studies were performed for different temperatures, particle sizes, pH and adsorbent amount. The removal was inversely proportional to the particle size of the fly ash. The results indicated that the Freundlich isotherm fitted better to the data than the Langmuir isotherm. Furthermore, the data correlated better with non-linear than the linear form.

Gandhimathi et al. 2012 used coal bottom ash as adsorbent material for the study of the removal of the crystal violet in single and tertiary systems. The bottom ash is from the Neyveli Lignite Corporation Limited, Thermal Power plant located in Neyveli, Tamil Nadu, India. In the kinetic and equilibrium studies, the pseudo-second order and Freundlich models fit the experimental data, respectively.

The removal of crystal violet on fly ash was investigated (Çoruh et al. 2012). The effects of pH, stirring time, initial dye concentration, initial dye solution and the adsorbent amount were investigated. The results showed that fly ash was effective in removing crystal violet from aqueous solution. The experimental data were well fitted to the Freundlich equation and describe by pseudo second-order model.

4. CRYSTAL VIOLET REMOVAL BY NANOSILICA

4.1. Characterizations of Nanosilica

Nanosilica adsorbent was characterized by several techniques. Scanning electron microscopy analysis showed the presence of very small nanoparticles (< 100 nm), X-ray diffraction showed a broad peak at 22° (2θ), a characteristic of silica in amorphous form (Rovani et al. 2018a).

The spectrum of nanosilica obtained by Fourier-transform infrared spectroscopy analysis showed four main bands: around 450 cm^{-1} and around 800 cm^{-1} can be assigned as symmetric stretching of siloxane groups; around 1050 cm^{-1} an asymmetric stretching of siloxane groups; and around 950 cm^{-1} is related to the angular deformation Si-OH of the silanol group (Rovani et al. 2018a).

The results of determination of pH$_{PZC}$ are presented as pH$_{final}$ values of solutions after their equilibration with SiO$_2$NP as a function of pH$_{initial}$ values of solutions (Figure 4a). It can be seen that the point of zero charge value of nanosilica was about 5.3 (Figure 4b). In solutions with pH value below 5.3, the silica has a positive charge, and in solutions, with a pH value above 5.3, the silica has a negative charge.

In Figure 4(a) the final pH value a common plateau was obtained. According to Milonjic et al. 2007, silica exhibits amphoteric properties. The plateau of the Figure 4(a) corresponds to the pH range where the buffering effect of the SiO$_2$NP surface takes place.

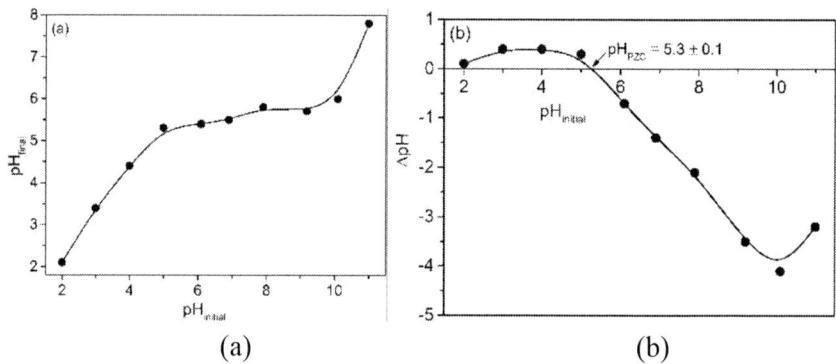

Figure 4. Determination of pH$_{PZC}$ (a) and pH$_{PZC}$ (b) of nanosilica obtained from sugarcane waste ash.

4.2. Adsorption Studies

The influence of the main operating parameters (initial pH, contact time and initial concentration of dye) was evaluated to achieve efficient removal of crystal violet using the SiO$_2$NP adsorbent.

Initially, adsorption tests were performed at different pH values between 2 and 10 with initial dye concentration of 24 mg L^{-1}. The adsorption capacity was maximal and remained stable at pH above 4 (Figure 5). The results obtained with solutions of crystal violet with pH values lower than 4 presented the unfavorable percentage of removal due to the high concentration of H$^+$ ions that compete with the dye cation at the adsorption sites of adsorbent. With the pH increases, more negatively charged are available resulting in decreased repulsion between the positively charged crystal violet molecule and the SiO$_2$NP. This result corroborates with the pH$_{PZC}$ value of the nanosilica shown in Figure 4.

In this way, kinetic study and equilibrium adsorption were performed without any adjustment of the pH of the crystal violet dye solution (pH between 5.0 and 6.0).

Figures 6a and 6b show the variation of the amount of crystal violet adsorbed (q$_t$) onto nanosilica as a function of time in two different initial concentrations 20 mg L^{-1} and 50 mg L^{-1}, respectively. It is possible to observe that crystal violet adsorption occurs very fast in the first minutes, and saturation was occurring after 60 min in both cases.

The dye adsorption mechanism can be attributed to the abundance of silanol (−OH) groups present in the nanosilica surface. In this case, the crystal violet molecules are likely to be attracted to the surface due to dipole-dipole interactions between the hydrogen of the surface of the adsorbent and the electropositive groups in the dye (Parker et al. 2012, Hung et al. 2018, Rovani et al. 2018a). As well as dipole-dipole H-bonding interaction between −OH groups present on the surface of the SiO$_2$NP and the electronegative residue (N lone pair) in the crystal violet molecule (Ghorai et al. 2014).

The values of the rate constant and adsorption capacity at equilibrium condition (q$_e$) of kinetic models (pseudo first-order and pseudo second-order) are presented in Table 8 along with the coefficient of determination values obtained from the contact time studies. The pseudo-second-order model yielded the highest coefficient of determination (R$^2_{adj.}$) values, which indicates that the experimental data are in good agreement with the pseudo second-order model.

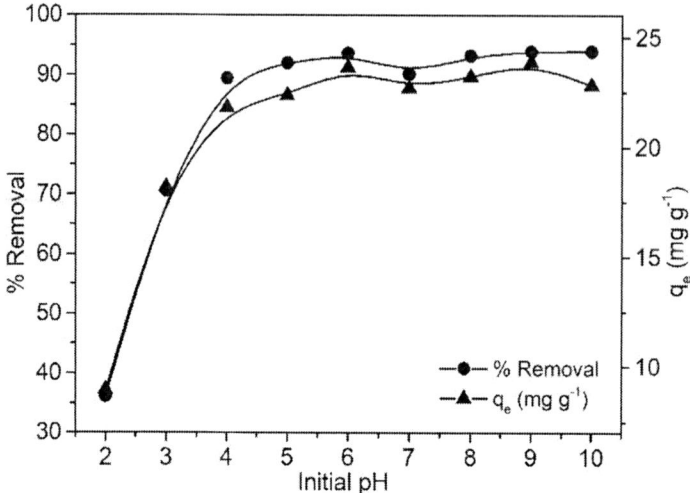

Figure 5. Removal of crystal violet dye by nanosilica adsorbent at different initial pH. Conditions: 25°C, initial concentration 24 mg L^{-1}, adsorbent dose 1.0 g L^{-1}, and contact time 24 h.

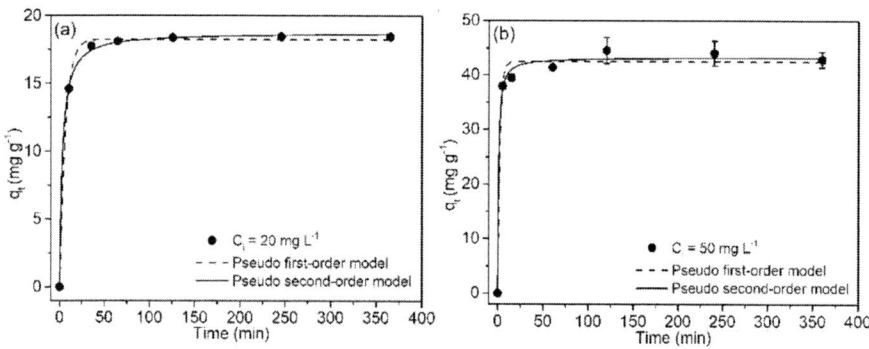

Figure 6. Pseudo first-order and pseudo second-order model kinetics plot for the removal of CV by nanosilica. Conditions: 25°C, adsorbent dose 1.0 g L^{-1}, and initial concentration 20 mg L^{-1} (a) and initial concentration 50 mg L^{-1} (b).

The pseudo second-order model suggests a mechanism of chemisorption, that is, can be the result of any shares or exchange of electrons between the crystal violet and the functional groups of the adsorbent (Blanchard et al. 1984; Ho et al. 1996; Eloussaief et al. 2011; Lima et al. 2015; Hamza et al. 2015, 2018; Asfaram et al. 2018).

Table 8. Kinetic parameters for crystal violet adsorption onto nanosilica

	Initial concentration of CV dye	
	[20 mg L^{-1}]	[50 mg L^{-1}]
Pseudo first-order		
k_f (min^{-1})	0.0087	0.0103
q_e (mg g^{-1})	18.24	42.52
R^2_{adj}	0.9985	0.9865
Pseudo second-order		
k_s (g mg^{-1} min^{-1})	0.0193	0.0284
q_e (mg g^{-1})	18.77	43.29
R^2_{adj}	0.9992	0.9936

Peres et al. 2018 evaluated the removal of the crystal violet dye from aqueous effluents using bio-nanosilica obtained from rice husk. For all concentrations studied, it was possible to remove 80% of the dye within 60 min. The pseudo second-order model was more adequate to represent the adsorption kinetics of bio-nanosilica.

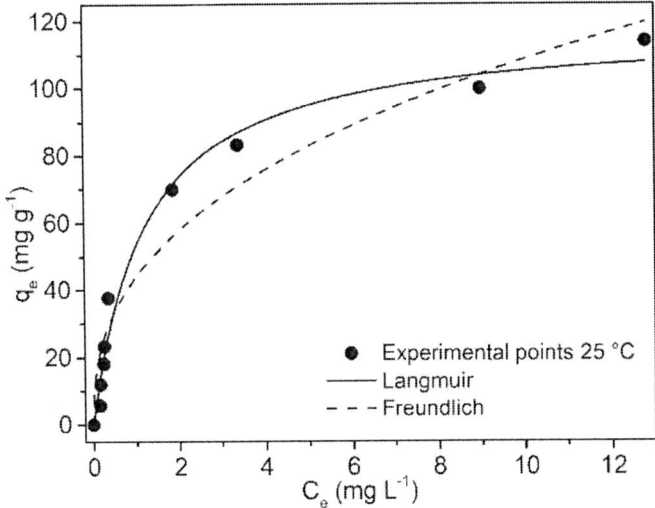

Figure 7. Adsorption Langmuir and Freundlich isotherms for crystal violet dye adsorbed by nanosilica. Conditions: 25°C, contact time 1 h and adsorbent dose 1.0 g L^{-1}.

Table 9. Parameters isotherm models for crystal violet dye adsorption onto nanosilica

Langmuir	
$Q^0_{max.}$ (mg g^{-1})	117.98
K_L (L mg^{-1})	0.824
R^2_{adj}	0.9805
χ^2	10.797
Freundlich	
K_F [(mg g^{-1}) (L mg^{-1})$^{1/n}$]	43.99
n_F	2.550
R^2_{adj}	0.9384
χ^2	3908.24

Hung et al. 2018 investigated the removal of crystal violet dye from aqueous solution by nanosilica from rice husk. The highest percentages of CV removal were obtained with a contact time of 60 min at pH 7, and the pseudo second-order model was the model that best predicted the experimental data.

The two most frequently isotherm models (Langmuir and Freundlich) were selected to describe the equilibrium data of CV adsorption onto nanosilica. The comparison of nonlinear fitted curves from experimental data is shown in Figure 7. The coefficients of determination ($R^2_{adj.}$), chi-square test (χ^2) and isotherm parameters from nonlinear regressive method were listed in Table 9.

The data fitted best with the Langmuir model which was predicted by the highest $R^2_{adj.}$ value and the lowest χ^2 value. Thus, adsorption equilibrium of crystal violet dye molecules by nanosilica occurs in monolayer in a finite and defined number of adsorption sites.

There are few studies on equilibrium isotherms of crystal violet onto adsorbents derived from nanosilica or sugarcane waste.

Allingham et al. 1958 presented isotherms of adsorption of cationic dyes by silica associated with the adsorption in monolayer, that is, Langmuir isotherm. The authors concluded that cationic dyes are adsorbed by ion exchange mainly as a monolayer of positively charged micelles on the external surface of the silica.

Adsorption studies of crystal violet from aqueous solution were carried out using the following adsorbents: sugarcane bagasse modified with EDTA dianhydride (Gusmão et al. 2013); nanocomposite of hydrolyzed polyacrylamide grafted xanthan gum/incorporated nanosilica (Ghorai et al. 2014); nanosilica obtained from rice husk (Hung et al. 2018); nanosilica-supported poly β-cyclodextrin (Chen et al. 2018). In all adsorption systems, Langmuir model fit the experimental data well, just like in the CV/nanosilica from sugarcane waste ash system described herein.

According to Jamwal et al. 2017, the Langmuir isotherm provides a more rational description of nanosilica adsorption as it considers the binding sites and their interactions, the surface homogeneity, and the power of the surface adsorbent.

CONCLUSION

The industrial activity, in the last decades, provided the economic growth, but also brought great environmental impacts. Among the different types of environmental pollution, water contamination is considered the most serious problem.

As mentioned in this chapter, crystal violet dye is used for many purposes and is known for its mutagenic and mitotic intoxication (Mani & Bharagava 2016; Kulkarni et al. 2017). As a result of the widespread use, a considerable amount of colored wastewater is generated and discharged into the environment.

One of the most important applications of crystal violet dye is in the textile industry in the dyeing process. This industrial sector is classified as one of the largest polluting sources of the environment in the world, considering the volume discharged and the composition of the effluent (Gümüş & Akbal 2011).

Effluents produced from the textile industries are a combination of various inorganic and organic compounds (Fulekar et al. 2013). The most of the dyes presenting organic groups toxic which can be they cause mortality, genotoxicity, mutagenicity and carcinogenicity effects in humans and animals (Kant 2012; Mani & Bharagava 2016).

The textile industries consume approximately 70% of the dyes (Roy et al. 2018). It is estimated that about 1 million tons of synthetic textile dyes are produced annually, of which 50,000 tons are lost to effluent during application and manufacturing, of which 5 - 15% are directly released into the environment without any suitable treatment (Olvera-Vargas et al. 2014; Rehman et al. 2017).

Currently, developing countries, e. g., China, India, Thailand, Turkey, use most of the dyes (Güyer et al. 2016; Cai et al. 2017). Many of these countries do not have strict legislation, or the discharge pattern is not enforced. Also, the high cost and technical difficulties make it even more difficult to eliminate pollution in the textile industries (Cai et al. 2017).

In Brazil, the National Council for the Environment (CONAMA) - Resolution Nº. 357 of March 17, 2005, establishes in the Articles 14 - 16 that in water bodies sweet of classes 1, 2 and 3, dyes from anthropic sources which are not removable by conventional coagulation, sedimentation, and filtration processes and in the Articles 18 - 20 that in saltwater bodies of classes 1, 2 and 3 dyes from anthropic sources must be virtually absent ("CONAMA. Resolution No. 357 of March 17, 2005. Amended by Resolution 410/2009 and 430/2011" 2011).

There are many studies in the literature reporting the removal of crystal violet from the aqueous medium. Several techniques have been applied using chemical, physical, and biological methods (Mondal 2008; Sharma & Kaur 2018). Among the physical methods for the removing dyestuffs in wastewater, the following are highlighted: adsorption, photo-oxidation, oxidation, advanced oxidation processes, filtration, coagulation, etc. (Salleh et al. 2011).

In recent years, adsorption has been extensively studied by researchers around the world as an efficient and economically sustainable technology to reduce the impact of pollution generated by the textile industry. It is considered a simple and low-cost method, presenting design flexibility, high selectivity and efficiency in removal (Chowdhury & Saha 2011; Chowdhury et al. 2013; Cai et al. 2017; Sharma & Kaur 2018).

A recent review article shows that the adsorption method with the use of low-cost adsorbent materials, such as agricultural materials and industrial waste, has been efficient in removing crystal violet dye (Sharma & Kaur 2018).

This chapter has demonstrated that modified zeolites of fly and bottom ashes from coal and silica from sugarcane waste ash may be an alternative for the removal of crystal violet. The adsorbent materials obtained in this study are low-cost and high value-added and can provide significant environmental and economic benefits.

In Brazil, about 3 to 12 million tons of ash from sugarcane waste ash per year are generated (CONAB 2018; Rovani et al. 2018b). In the case of coal ash, Brazilian and global thermoelectric plants generate around 8 and 800 million tons/year, respectively (Prado et al. 2017). Only a small part of the total production of these wastes is used. Faced with this scenario, the development of innovative recycling and environmentally friendly techniques is necessary.

Some suggestions can be taken from this chapter. Textile effluents can exhibit a multitude of substances different from one another, and in different concentrations. Thus, studies with real effluent containing crystal violet are recommended to evaluate the adsorption capacity of the adsorbents in a real condition.

Concerning the application in field tests, laboratory experiments should be extrapolated to pilot plant scale to produce larger quantities of the adsorbents. Cost studies of such a process should also be performed to assess commercial viability.

Effluents contaminated with dyes after treatment with zeolitic materials and nanosilica should not be disposed of directly into bodies of water but may be used as non-potable reuse water.

In Brazil, the National Water Resources Council (CNRH), resolution 54 of 2005 establishes modalities, guidelines, and criteria for the practice of water reuse for non-potable purposes. This reuse reduces the demand for water sources due to the replacement of drinking water by water of lesser quality. Non-potable reuse for industrial purposes can be applied in: processes, activities and industrial operations, dust abatement and cleaning in general ("CONAMA/CNRH. Resolution N° 54 of November 28, 2005" 2006).

One possibility to avoid the disposal of dye-saturated zeolite and nanosilica material is regeneration, which can be reused through the implementation of alternative technologies and that are economically accessible and environmentally correct (Jones et al. 2002; Ania et al. 2005; Yuen & Hameed 2009; Han et al. 2010; Rovani et al. 2018b). Future research on crystal violet is necessary for the development and implementation of new sustainable removal techniques.

ACKNOWLEDGMENTS

This study was financed in part by the Coordenação de Aperfeiçoamento de Pessoal de Nível Superior - Brasil (CAPES) - Finance Code 001, and in part by the Conselho Nacional de Desenvolvimento Científico e Tecnológico - Brazil (CNPq).

The authors also are grateful to Jorge Lacerda Thermoelectric Complex (Santa Catarina, Brazil) for providing coal ash samples and COSAN S.A. (São Paulo, Brazil) for supplying the sugarcane waste ash.

REFERENCES

Ahmad, R. (2009). Studies on adsorption of crystal violet dye from aqueous solution onto coniferous pinus bark powder (CPBP). *Journal of Hazardous Materials*, *171* (I), 767-773. doi: 10.1016/j.jhazmat.2009.06.060.

Alcantara, R. R., Izidoro, J. C. & Fungaro, D. A. (2015). Synthesis and characterization of surface modified nanomaterial zeolitic from coal fly ash. *International Journal of Materials Chemistry and Physics*, *1*, 370-377. Accessed November 12, 2018. https://pdfs.semanticscholar.org/3240/6068dcb9ffbdc229f10fc78663db3db3f4f8.pdf.

Alcantara, R. R., Muniz, R. O. R. & Fungaro, D. A. (2016). Full factorial experimental design analysis of Rhodamine B removal from water using organozeolite from coal bottom ash. *International Journal of Energy and Environment (Print)*, *7*, 357-374. Accessed November 12, 2018. http://www.ijee.ieefoundation.org/vol7/issue5/ IJEE_01_v7n5.pdf.

Allingham, M. M., Cullen, J. M., Giles, C. H., Jain, S. K. & Woods, J. S. (1958). Adsorption at inorganic surfaces. II. Adsorption of dyes and related compounds by silica. *Journal of Applied Chemistry*, *8* (II), 108-116. doi: 10.1002/jctb.5010080205.

Ali, A. Al-Haj. & El-Bishtawi, R. (1997). Removal of lead and nickel ions using zeolite tuff. *Journal of Chemical Technology & Biotechnology*, *69*, 27-34. doi: 10.1002/(SICI)1097-4660(199705)69:1<27::AID-JCTB682>3.0.CO;2-J.

Alves, R. H., Reis, T. V. D. S., Rovani, S. & Fungaro, D. A. (2017). Green synthesis and characterization of biosilica produced from sugarcane waste ash. *Journal of Chemistry*, 2017, 1-9. doi: 10.1155/2017/6129035.

Andreou, M. A. & Pashalidis, I. (2018). Removal of crystal violet from aqueous solution by biofibers. *Desalination and Water Treatment*, *112*, 90-93. doi: 10.5004/dwt.2018.22196.

Ania, C. O., Parra, J. B., Menéndez, J. A. & Pis, J. J. (2005). Effect of microwave and conventional regeneration on the microporous and mesoporous network and on the adsorptive capacity of activated carbons. *Microporous and Mesoporous Materials*, *85*, 7-15. doi: 10.1016/j.micromeso.2005.06.013.

Asfaram, A., Ghaedi, M., Dashtian, K. & Reza, G. G. (2018). Preparation and characterization of $Mn_{0.4}Zn_{0.6}Fe_2O_4$ nanoparticles supported on dead cells of Yarrowia lipolytica as a novel and efficient adsorbent/biosorbent composite for the removal of Azo food dyes: central composite design optimization study. *ACS Sustainable Chemistry Engineering*, *6*, 4549-4563. doi: 10.1021/acssuschemeng.7b03205.

Aysu, T. & Küçük, M. M. (2015). Removal of crystal violet and methylene blue from aqueous solutions by activated carbon prepared from ferula orientalis. *International Journal of Environmental Science and Technology*, *12* (VII), 2273-2284. doi: 10.1007/s13762-014-0623-y.

Bellir, K., Bouziane, I. S., Boutamine, Z., Lehocine, M. B. & Meniai, A. H. (2012). Sorption study of a basic dye 'gentian violet' from aqueous solutions using activated bentonite. *Energy Procedia*, *18*, 924-933. doi: 10.1016/j.egypro.2012.05.107.

Bertolini, T. C. R., Izidoro, J. C., Alcântara, R. R., Grosche, L. C. & Fungaro, D. A. (2015). Surfactant-modified zeolites from coal fly and bottom ashes as adsorbents for removal of crystal violet from aqueous solution. *Academica Editores: Acta Velit*, *2* (II), 1-17. Accessed November 12, 2018. http://saepub.com/acc.php?journal_name= ACTAVELIT&volume=2&issue=2.

Bertolini, T. C. R., Izidoro, J. C., Magdalena, C. P. & Fungaro, D. A. (2013). Adsorption of crystal violet dye from aqueous solution onto zeolites from coal fly and bottom ashes. *Orbital: The Electronic Journal of Chemistry*, *5* (III), 1-13. Accessed November 12, 2018. http://www.orbital.ufms.br/index.php/Chemistry/article/view/488.

Blanchard, G., Maunaye, M. & Martin, G. (1984). Removal of heavy metals from waters by means of natural zeolites. *Water Research*, *18* (XII), 1501-1507. doi: 10.1016/0043-1354(84)90124-6.

Cai, Zhengqing, Sun, Y., Liu, W., Pan, F., Sun, P. & Fu, J. (2017). An overview of nanomaterials applied for removing dyes from wastewater. *Environmental Science and Pollution Research*, *24* (XIX), 15882-15904. doi: 10.1007/s11356-017-9003-8.

Caro, H. & Kern, A. (1883). *Manufacture of dye-stuff*. United States Patent Office, Letters Patent No. 290856.

Carvalho, T. E. M., Fungaro, D. A., Magdalena, C. P. & Cunico, P. (2011). Adsorption of indigo carmine from aqueous solution using coal fly ash and zeolite from fly ash. *Journal of Radioanalytical and Nuclear Chemistry (Print)*, *289*, 617-626. doi: 10.1007/s10967-011-1125-8.

Chen, J., Pu, Y., Wang, C., Han, J., Zhong, Y. & Liu, K. (2018). Synthesis of a novel nanosilica-supported poly β-cyclodextrin sorbent and its properties for the removal of dyes from aqueous solution. *Colloids and Surfaces A: Physicochemical and Engineering Aspects*, *538*, 808-817. doi: 10.1016/j.colsurfa.2017.11.048.

Chien, S. H. & Clayton, W. R. (1980). Application of elovich equation to the kinetics of phosphate release and sorption in soils. *Soil Science Society of America Journal*, *44*, 265-268. doi: 10.2136/sssaj1980. 03615995004400020013x.

Chowdhury, S., Chakraborty, S. & Saha, P. D. (2013). Removal of crystal violet from aqueous solution by adsorption onto eggshells: Equilibrium, kinetics, thermodynamics and artificial neural network modeling. *Waste and Biomass Valorization*, *4* (III), 655-664. doi: 10.1007/s12649-012-9139-1.

Chowdhury, S. & Saha, P. (2011). Adsorption kinetic modeling of safranin onto rice husk biomatrix using pseudo-first- and pseudo-second-order kinetic models: Comparison of linear and non-linear methods. *Clean (Weinh)*, *39* (III), 274-282. doi: 10.1002/clen.201000170.

CONAB. (2018). *Companhia Nacional de Abastecimento - Acompanhamento da Safra Brasileira: Cana-de-Açúcar: Safra 2017/2018 Quarto Levantamento*, Abril/2018. Brasília. Accessed August 6, 2018. https://www.conab.gov.br/info-agro/safras/cana. [*National Supply Company - Follow-up of the Brazilian Harvest: Sugarcane: 2017/2018 Harvest Room Survey*]

"CONAMA. Resolution N°. 357 of March 17, 2005. (2005). Amended by Resolution 410/2009 and 430/2011. 2011. *DOU n° 053*. Accessed September 12, 2018. http://www2.mma.gov.br/port/ conama/res/res05/ res35705.pdf.

"CONAMA/CNRH. Resolution N° 54 of November 28, 2005. (2006). *DOU*. Accessed September 12, 2018. http://www.ceivap.org.br/ ligislacao/Resolucoes-CNRH/Resolucao-CNRH 54.pdf.

Cunico, P., Kumar, A. & Fungaro, D. A. (2015). Adsorption of dyes from simulated textile wastewater onto modified nano-zeolite from coal fly ash. *Journal of Nanoscience and Nanotechnology (Print)*, *1*, 148-161.

Cunico, P., Kumar, A., Alcantara, R. R. & Fungaro, D. A. (2018). Adsorption of solophenyl dyes from aqueous solution by modified nanozeolite from bottom ash and its toxicity to *C. dubia*. *Current Nanomaterials*, 2, 95-103. doi: 10.2174/2405461503666180201152351.

Çoruh, S., Geyikçi, F. & Nuri, O. N. (2012). Dye removal from aqueous solution by adsorption onto fly ash. In *ATINER'S Conference Paper Series*, *1*, Athens, Greece.

Eloussaief, A., Kallel, N., Yaacoubi, A. & Benzina, N. (2011). "Mineralogical identification, spectroscopic characterization, and potential environmental use of natural clay materials on chromate removal from aqueous solutions." *Chemical Engineering Journal*, *168* (III), 1024-1031. doi: 10.1016/j.cej.2011.01.077.

El-Sayed, G. O. (2011). Removal of methylene blue and crystal violet from aqueous solutions by palm kernel fiber. *Desalination*, *272* (I), 225-232. doi: 10.1016/j.desal.2011.01.025.

Fabryanty, R., Valencia, C., Soetaredjo, F. E., Putro, J. N., Santoso, S. P., Kurniawan, A., Ju, Y. H. & Ismadji, S. (2017). Removal of crystal violet dye by adsorption using bentonite – alginate composite. *Journal of Environmental Chemical Engineering*, *5* (VI), 5677-5687. doi: 10.1016/j.jece.2017.10.057.

Foletto, E. L., Hoffmann, R., Hoffmann, R. S., Portugal, Jr. U. L. & Jahn, S. L. (2005). Aplicabilidade das cinzas da casca de arroz. *Química Nova*, *28* (VI), 1055-1060. doi: 10.1590/S0100-40422005000600021.

Freundlich, H. (1906). Adsorption in Solution. *Zeitschrift Für Physikalische Chemie*, *40*, 1361-1368. [*Journal of Physical Chemistry*]

Fulekar, M. H., Wadgaonkar, S. L. & Singh, A. (2013). Decolourization of dye compounds by selected bacterial strains isolated from dyestuff industrial area. *International Journal of Advancements in Research & Technology*, *2* (VII), 182-192. Accessed November 12, 2018. http://www.ijoart.org/docs/Decolourization-of-Dye-Compounds-by-Selected-Bacterial-Strains-isolated-from-Dyestuff-Industrial-Area.pdf.

Fungaro, D. A., Bruno, M. & Grosche, L. C. (2009). Adsorption and kinetic studies of methylene blue on zeolite synthesized from fly ash. *Desalination and Water Treatment (Print)*, *2*, 231-239. doi: 10.5004/dwt.2009.305.

Fungaro, D. A., Grosche, L. C., Pinheiro, A. S., Izidoro, J. C. & Borrely, S. I. (2010). Adsorption of methylene blue from aqueous solution on zeolitic material for color and toxicity removal. *Orbital: the Electronic Journal of Chemistry*, 2, 235-247. Accessed November 12, 2018. http://www.orbital.ufms.br/index.php/Chemistry/article/view/129.

Fungaro, D. A., Borrely, S. I. & Carvalho, T. E. M. (2013). Surfactant modified zeolite from cyclone ash as adsorbent for removal of reactive orange 16 from aqueous solution. *American Journal of Environmental Protection*, 1 (I), 1-9. doi: 10.12691/env-1-1-1.

Fungaro, D. A., Cunico, P., Bertolini, T. C. T. & Alcântara, R. R. (2016). Zeolitic nanomaterial from coal ash modified with cationic surfactant: environmental applications and ecotoxicity. In *Cationic Surfactants: Properties, Uses and Toxicity*, edited by L. Sanders. Nova Science Publishers.

Gandhimathi, R., Ramesh, S. T., Sindhu, V. & Nidheesh, P. V. (2012). Single and tertiary system dye removal from aqueous solution using bottom ash: Kinetic and isotherm studies. *Iranica Journal of Energy & Environment*, 3 (I), 52-62. doi: 10.5829/idosi.ijee.2012.03.01.0113.

Ghorai, S., Sarkar, A., Raoufi, M., Panda, A. B., Schönherr, H. & Pal, S. (2014). Enhanced removal of methylene blue and methyl violet dyes from aqueous solution using a nanocomposite of hydrolyzed polyacrylamide grafted xanthan gum and incorporated nanosilica. *ACS Applied Materials & Interfaces*, 6 (VII), 4766-4777. doi: 10.1021/am4055657.

Gümüş, D. & Akbal, F. (2011). Photocatalytic degradation of textile dye and wastewater. *Water Air & Soil Pollution*, 216 (I), 117-124. doi: 10.1007/s11270-010-0520-z.

Gusmão, K. A., Gurgel, L. V., Melo, T. M. & Gil, L. F. (2013). Adsorption studies of methylene blue and gentian violet on sugarcane bagasse modified with EDTA dianhydride (EDTAD) in aqueous solutions: kinetic and equilibrium aspects. *Journal of Environmental Management*, 118, 135-143. doi: 10.1016/j.jenvman.2013.01.017.

Güyer, G. T., Nadeem, K. & Dizge, N. (2016). Recycling of pad-batch washing textile wastewater through advanced oxidation processes and its reusability assessment for turkish textile industry. *Journal of Cleaner Production*, *139*, 488-494. doi: 10.1016/j.jclepro.2016.08.009.

Hamza, W., Chtara, C. & Benzina, M. (2015). Characterization and application of Fe and iso-Ti-pillared bentonite on retention of organic matter contained in wet industrial phosphoric acid (54%): kinetic study. *Research on Chemical Intermediates*, *41*, 6117-6140. doi: 10.1007/s11164-014-1726-2.

Han, R., Wang, Y., Sun, Q., Wang, L., Song, J., He, X. & Dou, C. (2010). Malachite green adsorption onto natural zeolite and reuse by microwave irradiation. *Journal of Hazardous Materials*, *175* (I–III), 1056-1061. doi: 10.1016/j.jhazmat.2009.10.118.

Hassan, A. F., Abdelghny, A. M., Elhadidy, H. & Youssef, A. M. (2014). Synthesis and characterization of high surface area nanosilica from rice husk ash by surfactant-free sol-gel method. *Journal of Sol-Gel Science and Technology*, *69* (III), 465-472. doi: 10.1007/s10971-013-3245-9.

Henmi, T. (1987). Synthesis of hydroxy-sodalite ('zeolite') from waste coal ash. *Soil Science and Plant Nutrition*, *33* (III), 517-521. doi: 10.1080/00380768.1987.10557599.

Ho, Y. S. (2004). Selection of optimum sorption isotherm. *Carbon*, *42* (X), 2115-2116. doi: 10.1016/j.carbon.2004.03.019.

Ho, Y. S. (2006). Review of second-order models for adsorption systems. *Journal of Hazardous Materials*, *136* (III), 681-689. doi: 10.1016/j.jhazmat.2005.12.043.

Ho, Y. S., Wase, D. A. J. & Forster, C. F. (1996). (School of Civil Engineering, Birmingham University, Edgbaston, Birmingham B15 2TT (United Kingdom)) Forster. Kinetic studies of competitive heavy metal adsorption by sphagnum moss peat. *Environmental Technology (United Kingdom)*.

Ho, Y. S. & McKay, G. (1998). Sorption of dye from aqueous solution by peat. *Chemical Engineering Journal*, *70* (II), 115-124. doi: 10.1016/S0923-0467(98)00076-1.

Hung, N. V. Tam, L. H., Khoa, V. N. D. & Nhan, H. T. C. (2018). Synthesis of nanosilica from rice husk and optimization of the removal of crystal violet dye from aqueous solution. *Vietnam Journal of*

Science and Technology, 56 (IA), 189-196. doi: 10.15625/2525-2518/56/1A/12522.

Iler, R. K. (1979). *The chemistry of silica: Solubility, polymerization, colloid and surface properties and biochemistry of silica*. John Wiley and Sons.

Izidoro, J. C., Fungaro, D. A., Santos, F. S. & Wang, S. (2012). Characteristics of brazilian coal fly ashes and their synthesized zeolites. *Fuel Processing Technology, 97*, 38-44. doi: 10.1016/j.fuproc.2012.01.009.

Izidoro, J. C., Fungaro, D. A., Abbott, J. E. & Wang, S. (2013). Synthesis of zeolites X and A from fly ashes for cadmium and zinc removal from aqueous solutions in single and binary ion systems. *Fuel (Guildford), 103*, 827-834. doi: 10.1016/j.fuel.2012.07.060.

Izidoro, J. C., Miranda, C. S., Guilhen, S. N., Wang, S. & Fungaro, D. A. (2018). Treatment of coal ash landfill leachate using zeolitic materials from coal combustion by-products. *Advanced Materials and Technologies Environmental Sciences, 2*, 189-198.

Jamwal, H. S., Kumari, S., Chauhan, G. S., Reddy, N. S. & Ahn, J. H. (2017). Silica-polymer hybrid materials as methylene blue adsorbents. *Journal of Environmental Chemical Engineering, 5*, 103-113. doi: 10.1016/j.jece.2016.11.029.

Jones, D. A., Lelyveld, T. P., Mavrofidis, S. D., Kingman, S. W. & Miles, N. J. (2002). Microwave heating applications in environmental engineering - A Review. *Resources, Conservation and Recycling, 34* (II), 75-90. doi: 10.1016/S0921-3449(01)00088-X.

Kalapathy, U., Proctor, A. & Shultz, J. (2000a). A simple method for production of pure silica from rice hull ash. *Bioresource Technology, 73* (III), 257-262. doi: 10.1016/S0960-8524(99)00127-3.

Kalapathy, U., Proctor, A. & Shultz, J. (2000b). Production and properties of flexible sodium silicate films from rice hull ash silica. *Bioresource Technology, 72* (II), 99-106. doi: 10.1016/S0960-8524(99)00112-1.

Kalapathy, U., Proctor, A. & Shultz, J. (2002). An improved method for production of silica from rice hull ash. *Bioresource Technology, 85* (III), 285-289. doi: 10.1016/S0960-8524(02)00116-5.

Kamath, S. R. & Proctor, A. (1998). Silica gel from rice hull ash: preparation and characterization. *Cereal Chemistry, 75* (IV), 484-487. doi: 10.1094/CCHEM.1998.75.4.484.

Kant, R. (2012). Textile dyeing industry an environmental hazard. *Natural Science, 04* (I), 22-26. doi: 10.4236/ns.2012.41004.

Kulkarni, M. R., Revanth, T., Acharya, A. & Bhat, P. (2017). Removal of crystal violet dye from aqueous solution using water hyacinth: Equilibrium, kinetics and thermodynamics study. *Resource-Efficient Technologies, 3* (I), 71-77. doi: 10.1016/j.reffit.2017.01.009.

Kumar, K. V. (2006). Linear and non-linear regression analysis for the sorption kinetics of methylene blue onto activated carbon. *Journal of Hazardous Materials, 137* (III), 1538-1544. doi: 10.1016/j.jhazmat.2006.04.036.

Lagergren, S. (1898). Zur Theorie Der Sogenannten Adsorption Geloster Stoffe. *Kungliga Svenska Vetenskapsakademiens Handlingar, 24*, 1-39. [*Royal Swedish Academy of Sciences Documents*]

Langmuir, I. (1918). The adsorption of gases on plane surfaces of glass, mica and platinum. *Journal of the American Chemical Society, 40* (IX), 1361-1303. doi: 10.1021/ja02242a004.

Laskar, N. & Kumar, U. (2018). Adsorption of crystal violet from wastewater by modified bambusa tulda. *KSCE Journal of Civil Engineering, 22* (VIII), 2755-2763. doi: 10.1007/s12205-017-0473-5.

Li, Z. & Bowman, R. S. (1998). Sorption of perchloroethylene by surfactant-modified zeolite as controlled by surfactant loading. *Environmental Science & Technology, 32* (XV), 2278–82. doi: 10.1021/es971118r.

Lima, É. C., Adebayo, M. A. & Machado, F. M. (2015). Kinetic and equilibrium models of adsorption. In *Carbon Nanomaterials as Adsorbents for Environmental and Biological Applications*, edited by Carlos P Bergmann and Fernando Machado Machado, *33*. Cham: Springer International Publishing. doi: 10.1007/978-3-319-18875-1_3.

Magdalena, C. P. & Fungaro, D. A. (2014). Studies on removal of Acid Orange 8 from aqueous solution using HDTMA-modified zeolite from coal bottom ash. *International Journal of Advanced Research in*

Chemical Science, 1 (VII), 23-33. Accessed November 12, 2018. https://www.arcjournals.org/international-journal-of-advanced-research-in-chemical-science/volume-1-issue-7/4.

Mahmood, T., Saddique, M. T., Naeem, A., Westerhoff, P., Mustafa, S. & Alum, A. (2011). Comparison of different methods for the point of zero charge determination of NiO. *Industrial & Engineering Chemistry Research, 50,* 10017-10023. doi: 10.1021/ie200271d.

Mani, S. & Bharagava, R. N. (2016). Exposure to crystal violet, its toxic, genotoxic and carcinogenic effects on environment and its degradation and detoxification for environmental safety. In *Reviews of Environmental Contamination and Toxicology, 237,* edited by W P de Voogt, 71-104. Cham: Springer International Publishing. doi: 10.1007/978-3-319-23573-8_4.

Mclintock, I. S. (1967). The Elovich equation in chemisorption kinetics. *Nature, 216,* 1204-1205.

Milonjic, S. K., Cerovic, L. S., Cokesa, D. M. & Zec, S. (2007). The influence of cationic impurities in silica on its crystallization and point of zero charge. *Journal of Colloid and Interface Science, 309,* 155-159. doi: 10.1016/j.jcis.2006.12.033.

Miyah, Y., Lahrichi, A., Idrissi, M., Boujraf, S., Taouda, H. & Zerrouq, F. (2017). Assessment of adsorption kinetics for removal potential of crystal violet dye from aqueous solutions using moroccan pyrophyllite. *Journal of the Association of Arab Universities for Basic and Applied Sciences, 23* (I), 20-28. doi: 10.1016/j.jaubas.2016.06.001.

Mohan, D., Singh, K. P., Singh, G. & Kumar, K. (2002). Removal of dyes from wastewater using flyash, a low-cost adsorbent. *Industrial & Engineering Chemistry Research, 41* (XV), 3688-3895. doi: 10.1021/ie010667+.

Mondal, S. (2008). Methods of dye removal from dye house effluent-An Overview. *Environmental Engineering Science, 25* (III), 383-396. doi: 10.1089/ees.2007.0049.

Olvera-Vargas, H., Oturan, N., Aravindakumar, C. T., Paul, M. M. S., Sharma, V. K. & Oturan, M. A. (2014). Electro-oxidation of the dye azure b: kinetics, mechanism and by-products. *Environmental Science and Pollution Research International*, 21, 8379-8386. doi: 10.1007/s11356-014-2772-4.

Omer, O. S., Hussein, M. A., Hussein, B. H. M. & Mgaidi, A. (2018). Adsorption thermodynamics of cationic dyes (methylene blue and crystal violet) to a natural clay mineral from aqueous solution between 293.15 and 323.15 K. *Arabian Journal of Chemistry*, *11* (V), 615-623. doi: 10.1016/j.arabjc.2017.10.007.

Parker, H. L., Hunt, A. J., Budarin, V. L., Shuttleworth, P. S., Miller, K. L. & Clark, J. H. (2012). The importance of being porous: polysaccharide-derived mesoporous materials for use in dye adsorption. *RSC Advances*, *2*, 8992-8997. doi: 10.1039/C2RA21367B.

Peres, E. C., Favarin, N., Slaviero, J., Almeida, A. R. F., Enders, M. P., Muller, E. I. & Dotto, G. L. (2018). Bio-nanosilica obtained from rice husk using ultrasound and its potential for dye removal. *Materials Letters*, *231*, 72-75. doi: 10.1016/j.matlet.2018.08.018.

Prado, P. F., Nascimento, M., Yokoyama, L. & Cunha, O. G. C. (2017). Use of coal ash in zeolite production and applicationsin manganese adsorption. *American Journal of Engineering Research (AJER)*, *6* (XII), 394-403. Accessed November 12, 2018. http://www.ajer.org/papers/v6(12)/ZZF0612394403.pdf.

Rafiee, E., Shahebrahimi, S., Feyzi, M. & Shaterzadeh, M. (2012). Optimization of synthesis and characterization of nanosilica produced from rice husk (a common waste material). *International Nano Letters*, *2* (I), 29. doi: 10.1186/2228-5326-2-29.

Rehman, F., Sayed, M., Khan, J. A. & Khan, H. M. (2017). Removal of crystal violet dye from aqueous solution by gamma irradiation. *Journal of the Chilean Chemical Society*, *62*, 3359-3364. doi: 10.4067/S0717-97072017000100011.

Resende, N. G. A. M., Monte, M. B. M. & Paiva, P. R. P. (2008). Capítulo 39 - Zeolitas Naturais In *Rochas e Minerais Industriais – CETEM*, 2ª Edição, 889-15. Accessed November 12, 2018. http://minerais.cetem.gov.br/bitstream/cetem/1143/1/39.%20ZE%C3%93LITAS%20NATURAIS%20REVISADO.pdf.

Rovani, S., Santos, J. J., Corio, C. & Fungaro, D. A. (2018ª). Low cost silica nanoparticles biosorbent obtained from sugarcane waste ash: characterization and adsorption study of methylene blue dye. In *12° Encontro Brasileiro Sobre Adsorção*, 1-8. Gramado-RS. Accessed November 12, 2018. http://scheneventos.com.br/eba/envio/files/ 139_arq1.pdf. [*12th Brazilian Meeting on Adsorption*]

Rovani, S., Santos, J. J., Corio, P. & Fungaro, D. A. (2018b). Highly pure silica nanoparticles with high adsorption capacity obtained from sugarcane waste ash. *ACS Omega, 3* (III), 2618-2627. doi: 10.1021/acsomega.8b00092.

Roy, D. C., Biswas, S. K., Saha, A. K., Sikdar, B., Rahman, M., Roy, A. C., Prodhan, Z. H. & Tang, S. S. (2018). Biodegradation of crystal violet dye by bacteria isolated from textile industry effluents. *Peer J, 6,* 1-15. doi: 10.7717/peerj.5015.

Salleh, M. A. M., Mahmoud, D. K., Karim, W. A. W. A. & Idris, A. (2011). Cationic and anionic dye adsorption by agricultural solid wastes: a comprehensive review. *Desalination, 280* (I), 1-13. doi: 10.1016/j.desal.2011.07.019.

Sharma, S. & Kaur, A. (2018). Various methods for removal of dyes from industrial effluents - A Review. *Indian Journal of Science and Technology, 11* (XII), 1-21. doi: 10.17485/ijst/2018/v11i12/120847.

Tahir, N., Bhatti, H. N., Iqbal, M. & Noreen, S. (2017). Biopolymers composites with peanut hull waste biomass and application for crystal violet adsorption. *International Journal of Biological Macromolecules, 94,* 210-220. doi: 10.1016/j.ijbiomac.2016.10.013.

Tran, H. N., You, S. J., Hosseini-Bandegharaei, A. & Chao, H. P. (2017). Mistakes and inconsistencies regarding adsorption of contaminants from aqueous solutions: A critical review. *Water Research, 120,* 88–116. doi: 10.1016/j.watres.2017.04.014.

Yuen, F. K. & Hameed, B. H. (2009). Recent developments in the preparation and regeneration of activated carbons by microwaves. *Advances in Colloid and Interface Science*, *149* (I–II), 19-27. doi: 10.1016/j.cis.2008.12.005.

Zou, W., Zhao, L. & Han, R. (2011). Adsorption Characteristics of uranyl ions by manganese oxide coated sand in batch mode. *Journal of Radioanalytical and Nuclear Chemistry*, *288* (I), 239-249. doi: 10.1007/s10967-010-0904-y.

INDEX

A

acid, 13, 17, 18, 19, 35, 46, 51, 56, 61, 65, 71, 72, 73, 74, 76, 80, 81, 108
acidic, 14, 18, 52, 60
activated carbon, 33, 34, 42, 49, 50, 51, 61, 62, 66, 67, 68, 103, 104, 110, 114
adsorbent, v, viii, ix, 32, 35, 38, 42, 43, 44, 45, 46, 51, 53, 54, 59, 63, 65, 70, 71, 72, 74, 75, 76, 78, 82, 84, 86, 88, 89, 93, 94, 95, 96, 97, 99, 101, 103, 107, 111
adsorbent materials, v, ix, 75, 76, 78, 101
adsorption, vii, viii, ix, 5, 20, 32, 33, 34, 35, 38, 43, 44, 45, 46, 47, 48, 49, 50, 51, 52, 53, 60, 61, 62, 63, 64, 65, 66, 67, 68, 69, 70, 71, 72, 73, 74, 76, 78, 79, 80, 82, 83, 84, 85, 86, 88, 89, 90, 91, 92, 93, 94, 95, 97, 98, 99, 100, 101, 102, 103, 104, 105, 106, 107, 108, 110, 111, 112, 113, 114
adsorption isotherms, 82, 85
aggregation, 2, 3, 4, 8, 15
algae, 34, 52, 60
alginate, 34, 44, 46, 47, 48, 52, 54, 55, 56, 57, 58, 60, 63, 65, 66, 67, 70, 71, 79, 106
aquatic systems, ix, 75
aqueous solutions, 2, 3, 5, 9, 21, 49, 53, 54, 66, 68, 71, 72, 104, 106, 107, 109, 111, 113
aqueous suspension, 67, 72

B

bacteria, 34, 37, 40, 113
bacterial strains, 106
bioavailability, 60, 61, 78
biocompatibility, 52
biodegradability, 33, 52
biodegradation, viii, 32, 34, 40
biomass, viii, 32, 50, 62, 69, 70, 73, 113
biopolymer, 52, 62, 73
biosorbent, 73, 103, 113
by-products, 39, 109, 111

C

carbon, vii, 1, 13, 42, 68, 79, 108
carboxyl, 15, 52, 60, 69
cellulose, 34, 47, 52, 57, 62, 63, 68, 74
chemical, ix, 17, 18, 19, 34, 38, 42, 43, 45, 47, 51, 52, 55, 60, 61, 62, 76, 79, 100, 111

chemical bonds, 38
chemical industry, 52
chemical interaction, 60
chemisorption, 38, 45, 50, 53, 90, 96, 111
chitin, 34, 52, 53, 60, 69
chitosan, viii, 32, 34, 44, 47, 48, 52, 53, 55, 56, 58, 59, 60, 63, 67, 69, 70, 71, 72, 73, 74, 79
coal, vii, ix, 50, 65, 75, 76, 78, 80, 82, 86, 91, 92, 93, 101, 102, 103, 104, 105, 107, 108, 109, 110, 112
coal ashes, 76, 80, 86
complexation, 45, 46, 55
compounds, 4, 9, 10, 14, 17, 33, 39, 63, 67, 103, 106
contact time, ix, 45, 76, 82, 84, 85, 88, 94, 95, 96, 97, 98
copolymer, 13, 17, 18, 19, 54, 67
cost, ix, 34, 37, 38, 39, 41, 43, 56, 57, 60, 61, 62, 63, 64, 65, 66, 67, 68, 71, 73, 76, 78, 93, 100, 101, 111, 113
crystal violet, v, vii, viii, ix, 1, 2, 3, 11, 13, 14, 15, 17, 19, 20, 22, 24, 25, 31, 32, 33, 35, 36, 63, 65, 67, 68, 69, 70, 71, 72, 73, 74, 75, 76, 77, 78, 79, 80, 82, 83, 84, 85, 86, 91, 93, 94, 95, 96, 97, 98, 99, 100, 101, 102, 103, 104, 105, 106, 108, 110, 111, 112, 113
crystalline, 78, 79, 88

D

decontamination, 44, 49, 54, 63
deformation, 94
degradation, 33, 38, 39, 67, 70, 73, 107, 111
derivatives, 34, 53, 62, 63, 67, 68
desorption, 43, 45, 61, 71, 85
developing countries, 33, 100
dielectric constant, 3, 5, 15, 16
dyeing, viii, 2, 16, 17, 20, 21, 24, 25, 31, 33, 76, 77, 99, 110

dyes, 21, 33, 34, 35, 36, 38, 39, 42, 43, 44, 45, 47, 48, 52, 53, 54, 55, 60, 61, 62, 63, 66, 67, 68, 70, 71, 72, 74, 76, 78, 98, 99, 100, 101, 103, 104, 105, 107, 111, 112, 113

E

economic growth, 99
effluent, ix, 42, 64, 72, 76, 78, 93, 99, 100, 101, 111
electricity, 39, 42
electron, 9, 12, 13, 93
electron microscopy, 93
electrons, 2, 4, 10, 12, 13, 14, 46, 96
energy, 4, 5, 6, 12, 13, 34, 39
environment, ix, 33, 34, 36, 37, 39, 40, 63, 64, 70, 75, 78, 99, 100, 111
environmental impact, 99
equilibrium, ix, 2, 5, 6, 49, 72, 73, 76, 81, 83, 84, 85, 86, 88, 89, 90, 92, 93, 95, 98, 107, 110
ethylene glycol, 47, 55, 59

F

fibers, vii, 2, 17, 20, 21, 22, 24, 62
filtration, 34, 41, 78, 100
flexibility, 3, 10, 38, 100
flocculation, 34, 39, 69, 78
flora and fauna, 36
food, viii, 20, 32, 37, 55, 63, 76, 103
food chain, viii, 32
food industry, 37, 63
formation, 2, 4, 9, 38, 53, 86, 88
functionalization, 60, 62

G

geometry, 11, 13, 25

Index

ginger, 46, 49, 56, 69
graphite, 48, 53, 71

H

hazardous materials, 68
heavy metals, 38, 70, 104
hydrogen, 14, 39, 45, 46, 47, 48, 95
hydrophilic, 2, 11, 24
hydrophobic, 2, 4, 5, 7, 8, 11, 15, 20, 24, 46, 47, 78
hypsochromic effect, vii, 1, 2

I

industries, ix, 34, 36, 63, 75, 76, 99, 100
industry, 37, 61, 76, 99, 100, 108, 110, 113
ions, 2, 3, 5, 9, 11, 14, 15, 41, 53, 55, 95, 103, 114
irradiation, 34, 71, 108, 112
isotherms, 45, 70, 85, 90, 97, 98

J

jaundice, viii, 32

K

kidney failure, 37
kinetic constants, 89
kinetic model, 89, 95, 105
kinetic parameters, 90
kinetic studies, 65, 69, 70, 88, 106
kinetics, 45, 49, 60, 65, 68, 71, 72, 73, 83, 88, 89, 96, 97, 105, 110, 111

L

light, vii, viii, 1, 11, 13, 32, 33, 37
light transmission, 33

M

macromolecular chains, 21
mammalian cells, ix, 75, 77
manufacturing, ix, 36, 75, 100
materials, ix, 16, 34, 45, 47, 48, 50, 52, 55, 61, 62, 63, 70, 71, 76, 78, 81, 86, 87, 88, 101, 106, 109
medicine, viii, 32, 77
mesoporous materials, 112
metal ion, 45, 52, 53, 65
methylene blue, 65, 68, 70, 71, 76, 104, 106, 107, 109, 110, 112, 113
microorganisms, 18, 39, 40
models, ix, 46, 76, 84, 85, 90, 92, 93, 98, 105, 108, 110
modifications, 35, 43, 47, 64
molecular mass, 18, 19
molecular structure, 9, 35, 38, 61
molecular weight, 41, 59
molecules, 3, 4, 5, 6, 7, 8, 12, 14, 16, 24, 25, 47, 53, 60, 88, 90, 95, 98
monolayer, 38, 86, 87, 90, 98

N

nanomaterials, 104
nanoparticles, 49, 50, 68, 72, 73, 82, 93, 103, 113
nanosilica, vii, ix, 75, 76, 79, 80, 81, 93, 94, 95, 96, 97, 98, 99, 101, 102, 104, 107, 108, 112

O

optimization, 60, 65, 72, 73, 103, 108
organic compounds, 26, 35, 41, 78, 99
oxidation, 34, 39, 71, 78, 100, 107, 111

P

pH, viii, 10, 14, 32, 35, 40, 44, 47, 50, 51, 53, 56, 57, 58, 59, 60, 77, 80, 81, 87, 93, 94, 95, 96, 98
pharmaceutical, 63
physico-chemical, 43, 45
physicochemical characteristics, 86
physicochemical properties, ix, 76
physisorption, 38, 45
pollutants, 33, 38, 43, 45, 52, 53, 54, 60, 61, 63
pollution, 32, 33, 37, 99, 100
polyacrylamide, 79, 99, 107
polymer, 2, 7, 8, 10, 14, 17, 44, 52, 57, 60, 69, 109
polymer chain, 7, 10
polymer molecule, 7, 14
polymerization, 5, 7, 10, 11, 44, 109
polysaccharide, 45, 54, 60, 73, 112
polysaccharides, 34, 52
polyurethane, 48, 53, 71
precipitation, 45, 46, 78, 80
preparation, iv, 33, 37, 44, 55, 110, 114

R

radical polymerization, 17
recovery, 41, 43, 44, 54, 62, 68
regeneration, 42, 61, 66, 102, 103, 114
regression, 46, 71, 85, 92, 110
regression analysis, 110
regression method, 46, 85
remission, v, vii, 1, 2, 22, 24, 25
requirements, 34, 40, 52, 63
researchers, 33, 43, 52, 54, 55, 79, 100
rice husk, viii, 32, 50, 65, 79, 97, 98, 99, 105, 108, 112

S

silica, 78, 79, 82, 93, 94, 98, 101, 103, 109, 111, 113
sodium, 11, 13, 14, 17, 18, 19, 50, 55, 58, 65, 80, 81, 109
sodium dodecyl sulfate, 50
sodium hydroxide, 81
solid phase, 45, 78
solid waste, 72, 113
solution, viii, 2, 3, 4, 5, 6, 7, 11, 14, 15, 21, 25, 32, 35, 45, 60, 61, 64, 65, 66, 67, 68, 69, 70, 71, 72, 73, 80, 81, 82, 83, 84, 86, 87, 93, 95, 98, 99, 102, 103, 104, 105, 106, 107, 108, 110, 112
sorption, 38, 45, 46, 47, 54, 55, 60, 61, 63, 67, 73, 105, 108, 110
structure, viii, 3, 9, 12, 17, 18, 19, 32, 34, 35, 36, 43, 44, 45, 46, 52, 54, 77, 78, 88
styrene, 12, 13, 17, 18, 19, 24, 25, 46, 57, 66, 67
sugarcane, vii, viii, ix, 32, 62, 75, 78, 79, 80, 81, 94, 98, 99, 101, 102, 103, 107, 113
sugarcane waste ash, vii, ix, 75, 78, 80, 81, 94, 99, 101, 102, 103, 113
surface area, 43, 61, 87, 92, 108
surface modification, 47, 68
surfactant, vii, viii, ix, 32, 65, 70, 71, 75, 80, 82, 86, 87, 88, 90, 107, 108, 110
surfactant-modified zeolite, vii, ix, 75, 80, 82, 86, 87, 88, 104, 110
synthesis, 62, 64, 73, 80, 88, 103, 112

T

techniques, 33, 34, 38, 78, 93, 100, 101, 102
temperature, viii, 16, 20, 32, 35, 49, 53, 60, 74, 81
temperature dependence, 60
terrestrial ecosystems, 63

thermodynamics, 49, 65, 68, 71, 105, 110, 112
toxicity, viii, 32, 33, 34, 36, 37, 52, 76, 105, 107
treatment, viii, 32, 33, 34, 36, 37, 38, 39, 40, 43, 44, 50, 52, 61, 62, 64, 66, 67, 72, 80, 86, 87, 100, 101
treatment methods, viii, 32, 34, 37, 38, 39
treatment of effluents, 78

X

xanthan gum, 79, 99, 107

Z

zeolite, vii, ix, 50, 75, 76, 80, 82, 86, 87, 88, 90, 91, 102, 103, 104, 105, 106, 107, 108, 110, 112

V

vomiting, viii, 32

W

wastewater, ix, 33, 34, 36, 38, 41, 43, 52, 63, 64, 65, 66, 67, 68, 69, 70, 72, 75, 99, 100, 104, 105, 107, 110, 111
water, vii, viii, ix, 10, 11, 14, 15, 16, 20, 21, 32, 33, 36, 37, 49, 50, 55, 57, 60, 61, 63, 66, 67, 68, 70, 75, 78, 81, 82, 87, 88, 99, 100, 101, 103, 110

Related Nova Publications

COAL FLY ASH: PROPERTIES, APPLICATIONS AND PERFORMANCE

EDITOR: Miriam Gray

SERIES: Environmental Remediation Technologies, Regulations and Safety

BOOK DESCRIPTION: Due to rising energy demand, the amount of coal utilization is escalating at a fast rate, leading to a rise in ash generation. *Coal Fly Ash: Properties, Applications and Performance* begins by presenting a study wherein the physiochemical properties of coal ash were analysed and based on that, their potential for reuse was identified.

SOFTCOVER ISBN: 978-1-53614-511-3
RETAIL PRICE: $82

To see complete list of Nova publications, please visit our website at www.novapublishers.com